This revised edition of Joint Publication 3-0, *Joint Operations*, reflects the current guidance for conducting joint activities across the range of military operations and is the basis for US participation in multinational operations where the US has not ratified specific doctrine or procedures. This keystone publication forms the core of joint warfighting doctrine and establishes the framework for our forces' ability to fight as a joint team.

Often called the "linchpin" of the joint doctrine publication hierarchy, the overarching constructs and principles contained in this publication provide a common perspective from which to plan and execute joint operations independently or in cooperation with our multinational partners, other US Government departments and agencies, and international and nongovernmental organizations.

As our nation continues into the 21st century, the guidance in this publication will enable current and future leaders of the Armed Forces of the United States to design, plan, organize, train for, and execute worldwide missions as our forces transform to meet emerging challenges. To succeed, we need adaptive, agile, and timely doctrine for thinking professionals who understand the capabilities their Service brings to joint operations; how to integrate those capabilities with those of the other Services and interorganizational partners to optimize the strength of unified action; and how to organize, employ, and sustain joint forces to provide national leaders with multiple options for addressing various security threats. The focus is to improve joint warfighting, enhance readiness, and continue development of leaders for the joint force. Above all, we need professionals imbued with a sense of commitment and honor who will act decisively in the absence of specific guidance.

I ask all commanders to ensure the widest distribution of this keystone joint publication and actively promote the use of joint publications at every opportunity. I further ask you to study and understand the guidance contained in this publication and teach these principles to your subordinates. Only then will we be able to fully exploit the remarkable military potential inherent in our joint teams.

For the Chairman of the Joint Chiefs of Staff:

KEVIN D. SCOTT
Vice Admiral, USN
Director, Joint Force Development

PREFACE

1. Scope

This publication is the keystone document of the joint operations series. It provides the doctrinal foundation and fundamental principles that guide the Armed Forces of the United States in all joint operations.

2. Purpose

This publication has been prepared under the direction of the Chairman of the Joint Chiefs of Staff. It sets forth joint doctrine to govern the activities and performance of the Armed Forces of the United States in joint operations, and it provides considerations for military interaction with governmental and nongovernmental agencies, multinational forces, and other interorganizational partners. It provides military guidance for the exercise of authority by combatant commanders and other joint force commanders (JFCs), and prescribes joint doctrine for operations and training. It provides military guidance for use by the Armed Forces in preparing and executing their plans and orders. It is not the intent of this publication to restrict the authority of the JFC from organizing the force and executing the mission in a manner the JFC deems most appropriate to ensure unity of effort in the accomplishment of objectives.

3. Application

a. Joint doctrine established in this publication applies to the Joint Staff, commanders of combatant commands, subordinate unified commands, joint task forces, subordinate components of these commands, the Services, and combat support agencies.

b. The guidance in this publication is authoritative; as such, this doctrine will be followed except when, in the judgment of the commander, exceptional circumstances dictate otherwise. If conflicts arise between the contents of this publication and the contents of Service publications, this publication will take precedence unless the Chairman of the Joint Chiefs of Staff, normally in coordination with the other members of the Joint Chiefs of Staff, has provided more current and specific guidance. Commanders of forces operating as part of a multinational (alliance or coalition) military command should follow multinational doctrine and procedures ratified by the United States. For doctrine and procedures not ratified by the US, commanders should evaluate and follow the multinational command's doctrine and procedures, where applicable and consistent with US law, regulations, and doctrine.

Intentionally Blank

SUMMARY OF CHANGES
REVISION OF JOINT PUBLICATION 3-0
DATED 11 AUGUST 2011

- Revises Chapter V, "Joint Operations Across the Conflict Continuum," to expand the discussion of the changing balance of military activities in different types of military operations.

- Distributes pertinent information from Chapter V, "Joint Operations Across the Conflict Continuum," into Chapters VI, "Military Engagement, Security Cooperation, and Deterrence," VII, "Crisis Response and Limited Contingency Operations," and VIII, "Large-Scale Combat Operations" to enhance readability.

- Clarifies notional phasing model construct and associated graphics.

- Incorporates current information on joint electromagnetic spectrum management operations and protection of civilians.

- Reduces redundancies and improves continuity between Joint Publication (JP) 3-0, *Joint Operations;* JP 1, *Doctrine for the Armed Forces of the United States,* and JP 5-0; *Joint Planning.*

- Establishes continuity with new JP 3-20, *Security Cooperation.*

- Updates information on assessment.

- Updates terms and definition.

Intentionally Blank

TABLE OF CONTENTS

EXECUTIVE SUMMARY ix

CHAPTER I
FUNDAMENTALS OF JOINT OPERATIONS

- Introduction I-1
- Strategic Environment and National Security Challenges I-2
- Instruments of National Power and the Conflict Continuum I-4
- Strategic Direction I-5
- Unified Action I-8
- Levels of Warfare I-12
- Characterizing Military Operations and Activities I-14

CHAPTER II
THE ART OF JOINT COMMAND

- Introduction II-1
- Commander-Centric Leadership II-1
- Operational Art II-3
- Operational Design II-4
- Joint Planning II-5
- Assessment II-8

CHAPTER III
JOINT FUNCTIONS

- Introduction III-1
- Command and Control III-2
- Intelligence III-23
- Fires III-26
- Movement and Maneuver III-33
- Protection III-35
- Sustainment III-42

CHAPTER IV
ORGANIZING FOR JOINT OPERATIONS

- Introduction IV-1
- Understanding the Operational Environment IV-1
- Organizing the Joint Force IV-4
- Organizing the Joint Force Headquarters IV-8
- Organizing Operational Areas IV-10

Table of Contents

CHAPTER V
JOINT OPERATIONS ACROSS THE CONFLICT CONTINUUM

- Introduction V-1
- Military Operations and Related Missions, Tasks, and Actions V-1
- The Range of Military Operations V-4
- The Theater Campaign V-5
- A Joint Operation Model V-7
- Phasing a Joint Operation V-12
- The Balance of Offense, Defense, and Stability Activities V-15
- Linear and Nonlinear Operations V-17

CHAPTER VI
MILITARY ENGAGEMENT, SECURITY COOPERATION, AND DETERRENCE

- Introduction VI-1
- Typical Operations and Activities VI-5
- Other Considerations VI-12

CHAPTER VII
CRISIS RESPONSE AND LIMITED CONTINGENCY OPERATIONS

- Introduction VII-1
- Crisis Response and Limited Contingency Operations VII-1
- Typical Operations VII-2
- Other Considerations VII-6

CHAPTER VIII
LARGE-SCALE COMBAT OPERATIONS

- Introduction VIII-1
- Combatant Command Planning VIII-2
- Setting Conditions for Theater Operations VIII-4
- Considerations for Deterrence VIII-6
- Considerations for Seizing the Initiative VIII-12
- Considerations for Dominance VIII-19
- Considerations for Stabilization VIII-25
- Considerations for Enabling Civil Authority VIII-28

APPENDIX

    A   Principles of Joint Operations A-1
    B   References B-1
    C   Administrative Instructions C-1

GLOSSARY

Part I	Abbreviations, Acronyms, and Initialisms	GL-1
Part II	Terms and Definitions	GL-6

FIGURE

I-1	Principles of Joint Operations	I-2
I-2	Common Operating Precepts	I-3
I-3	Unified Action	I-9
I-4	Relationship Between Strategy and Operational Art	I-13
II-1	Joint Planning Process	II-6
II-2	Assessment Interaction	II-11
III-1	Command Relationships Synopsis	III-4
III-2	Building Shared Understanding	III-15
III-3	Risk Management Process	III-20
IV-1	A Systems Perspective of the Operational Environment	IV-3
IV-2	Notional Joint Force Headquarters and Cross-Functional Staff Organization	IV-9
IV-3	Operational Areas within a Theater	IV-11
IV-4	Contiguous and Noncontiguous Operational Areas	IV-12
V-1	Examples of Military Operations and Activities	V-2
V-2	Notional Operations Across the Conflict Continuum	V-4
V-3	Notional Joint Operations in a Theater Campaign Context	V-6
V-4	A Notional Joint Combat Operation Model	V-8
V-5	Notional Balance of Activities for a Joint Strike	V-11
V-6	Notional Balance of Activities for a Foreign Humanitarian Assistance Operation	V-12
V-7	Phasing an Operation Based on Predominant Military Activities	V-13
V-8	Notional Balance of Offense, Defense, and Stability Activities	V-15
V-9	Combinations of Areas of Operations and Linear/Nonlinear Operations	V-19
VI-1	The Conflict Continuum	VI-2

Intentionally Blank

EXECUTIVE SUMMARY
COMMANDER'S OVERVIEW

- **Presents the Fundamentals of Joint Operations**

- **Describes the Art of Joint Command**

- **Covers Joint Functions**

- **Explains Organizing for Joint Operations**

- **Characterizes Joint Operations across the Conflict Continuum**

- **Discusses Military Engagement, Security Cooperation, and Deterrence**

- **Addresses Crisis Response and Limited Contingency Operations**

- **Describes Large-Scale Combat Operations**

Fundamentals of Joint Operations

Joint operations are military actions conducted by joint forces and those Service forces employed in specified command relationships with each other, which of themselves do not establish joint forces. A joint force is one composed of significant elements, assigned or attached, of two or more Military Departments operating under a single joint force commander.

Joint Publication (JP) 3-0 is the keystone document in the joint operations series and is a companion to joint doctrine's capstone JP 1, *Doctrine for the Armed Forces of the United States*. It provides guidance to joint force commanders (JFCs) and their subordinates to plan, execute, and assess joint military operations. It also informs interagency and multinational partners, international organizations, nongovernmental organizations (NGOs), and other civilian decision makers of fundamental principles, precepts, and philosophies that guide the employment of the Armed Forces of the United States.

Principles of Joint Operations

Joint doctrine recognizes the nine principles of war (objective, offensive, mass, maneuver, economy of force, unity of command, security, surprise, and simplicity). Experience gained in a variety of irregular warfare situations has reinforced the value of three additional principles—restraint, perseverance, and legitimacy.

Strategic Environment and National Security Challenges

The military environment and the threats it presents are increasingly *transregional, multi-*

Executive Summary

domain, and *multi-functional* in nature. Today's potential adversaries can increasingly synchronize, integrate, and direct lethal operations and other non-lethal elements of power with greater sophistication, and are less constrained by geographic, functional, legal, or phasing boundaries. Conflict is now, and will remain, inherently transregional as future potential adversaries' interests, influence, capabilities, and reach extend beyond single areas of operation.

Instruments of National Power and the Conflict Continuum

US instruments of national power are the national-level means our national leaders can apply in various ways to achieve strategic objectives (ends). US national leaders can use military capabilities in a wide variety of activities, tasks, missions, and operations that vary in purpose, scale, risk, and combat intensity along the conflict continuum. The potential range of military activities and operations extends from military engagement, security cooperation, and deterrence in times of relative peace up through major operations and campaigns that typically involve large-scale combat.

Strategic Direction

National strategic direction is governed by the Constitution, federal law, United States Government policy, internationally recognized law, and the national interest as represented by national security policy.

In general, the President frames the strategic context by defining national interests and goals in documents such as the National Security Strategy (NSS), Presidential policy directives, executive orders, and other national strategic documents, in conjunction with the National Security Council and Homeland Security Council. The documents outline how the Department of Defense (DOD) will support NSS objectives and provide a framework for other DOD policy and planning guidance, such as the Guidance for Employment of the Force (GEF), Defense Planning Guidance, Global Force Management Guidance, and the Joint Strategic Capabilities Plan (JSCP).

Unified Action

Unified action is a comprehensive approach that focuses on coordination and cooperation of the US military and other interorganizational participants toward common objectives, even if the participants are not necessarily part of the same command or organization. This publication uses the term interorganizational participants to refer collectively to United States Government departments and agencies (i.e., interagency partners); state, territorial, local, and tribal agencies; foreign military forces and government agencies (i.e., multinational partners); NGOs; and the private sector. Joint forces must be prepared to plan and execute operations with forces from partner nations within the framework of an alliance or coalition under US or other-than-US leadership. By law, the President retains command authority over US forces.

Levels of Warfare

Three levels of warfare—strategic, operational, and tactical—model the relationship between national objectives and tactical actions. The operational level of warfare links the tactical employment of forces to national strategic objectives.

The Art of Joint Command

Introduction

Command is the authority that a commander in the armed forces lawfully exercises over subordinates by virtue of rank or assignment. While command authority stems from orders and other directives, the art of command resides in the commander's ability to use leadership to maximize performance.

Commander-Centric Leadership

Clear commander's guidance and intent, enriched by the commander's experience and intuition, enable joint forces to achieve objectives. The command and control (C2) function is commander-centric and network-enabled to facilitate initiative and decision making at the lowest appropriate level. If a commander loses reliable communications, **mission command**—a key component of the C2 [joint] function—enables military operations through decentralized execution based on mission-type orders. Commanders delegate decisions to subordinates wherever possible, which minimizes detailed control and empowers subordinates' initiative to make decisions based on the commander's guidance rather than constant communications.

Executive Summary

Operational Art

The commander is the central figure in operational art, not only due to education and experience, but also because the commander's judgment and decisions guide the staff throughout joint planning and execution.

Operational art is the cognitive approach by commanders and staffs–supported by their skill, knowledge, experience, creativity, and judgment– to develop strategies, campaigns, and operations to organize and employ military forces by integrating ends, ways, and means. The foundation of operational art encompasses broad vision; the ability to anticipate; and the skill to plan, prepare, execute, and assess. It helps commanders and their staffs organize their thoughts and envision the conditions necessary to accomplish the mission and reach the desired military end state in support of national objectives.

Operational Design

Operational design is the conception and construction of the framework that underpins a campaign or major operation plan and its subsequent execution. Operational design supports operational art with a methodology designed to enhance understanding the situation and the problem. Elements of operational design—such as objective, center of gravity (COG), line of operation (LOO), line of effort, and termination—are tools that help the JFC and the staff visualize and describe the broad operational approach to achieve objectives and accomplish the mission.

Joint Planning

Planning translates guidance into plans or orders to achieve a desired objective or attain an end state. The joint planning process aligns military activities and resources to achieve national objectives and enables leaders to examine cost-benefit relationships, risks, and trade-offs to determine a preferred course of action (COA) to achieve that objective or attain an end state.

Assessment

Assessment is a continuous process that measures the overall effectiveness of employing joint force capabilities during military operations. Theater-strategic and operational-level assessments provide a methodology for joint commands and Services to adjust planning and execution to be more effective, match the dynamic operational environment (OE), and better identify their risks and opportunities.

Executive Summary

Joint Functions

Introduction

Joint functions are related capabilities and activities grouped together to help JFCs integrate, synchronize, and direct joint operations. Functions that are common to joint operations at all levels of warfare fall into six basic groups—C2, intelligence, fires, movement and maneuver, protection, and sustainment.

Command and Control

C2 encompasses the exercise of authority and direction by a commander over assigned and attached forces to accomplish the mission. Command includes both the authority and responsibility to use resources to accomplish assigned missions. Control is inherent in command. To control is to manage and direct forces and functions consistent with a commander's command authority. Control provides the means for commanders to maintain freedom of action, delegate authority, direct operations from any location, and integrate and synchronize actions throughout the operational area (OA).

Intelligence

The intelligence function supports this understanding with analysis of the OE to inform JFCs about adversary capabilities, COGs, vulnerabilities, and future COAs and to help commanders and staffs understand and map friendly, neutral, and threat networks. Using the continuous joint intelligence preparation of the operational environment (JIPOE) analysis process, properly tailored JIPOE products can enhance OE understanding and enable the JFC to act inside the enemy's decision cycle.

Fires

To employ fires is to use available weapons and other systems to create a specific effect on a target. Joint fires are those delivered during the employment of forces from two or more components in coordinated action to produce desired results in support of a common objective.

Movement and Maneuver

This function encompasses the disposition of joint forces to conduct operations by securing positional advantages before or during combat operations and

Executive Summary

by exploiting tactical success to achieve operational and strategic objectives. Maneuver is the employment of forces in the OA through movement in combination with fires to achieve a position of advantage in respect to the enemy.

Protection

The protection function encompasses force protection, force health protection (FHP), and other protection activities. The function focuses on force protection, which preserves the joint force's fighting potential in four primary ways (active defense, passive defense, application of technology and procedures to reduce the risk of friendly fire incidents, and emergency management and response). FHP complements force protection efforts by promoting, improving, preserving, or restoring the mental or physical well-being of Service members. As the JFC's mission requires, the protection function also extends beyond force protection to encompass protection of US noncombatants.

Sustainment

Sustainment is the provision of logistics and personnel services to maintain operations through mission accomplishment and redeployment of the force. Sustainment provides the JFC the means to enable freedom of action and endurance and to extend operational reach. Sustainment determines the depth to which the joint force can conduct decisive operations, allowing the JFC to seize, retain, and exploit the initiative.

Organizing for Joint Operations

Understanding the Operational Environment

The JFC's OE is the composite of the conditions, circumstances, and influences that affect employment of capabilities and bear on the decisions of the commander. It encompasses physical areas of the air, land, maritime, and space domains; the information environment (which includes cyberspace); the electromagnetic spectrum; and other factors. Included within these are enemy, friendly, and neutral systems that are relevant to a specific joint operation. The nature and interaction of these systems will affect how the commander plans, organizes for, and conducts joint operations.

Executive Summary

Organizing the Joint Force

The JFC's mission and operational approach, as well as the principle of unity of command and a mission command philosophy, are guiding principles to organize the joint force for operations. Joint forces can be established on a geographic or functional basis. When JFCs organize their forces, they should also consider the degree of interoperability among Service components, with multinational forces and other potential participants. Joint force options include combatant commands (CCMDs), subordinate unified commands, and joint task forces. All JFCs may conduct operations through their Service component commanders, lower-echelon Service force commanders, and functional component commanders. Commander, US Special Operations Command, exercises combatant command (command authority) of all special operations forces (SOF). Geographic combatant commanders (GCCs) exercise operational control (OPCON) of their supporting theater special operations commands and most often exercise OPCON of SOF deployed in their areas of responsibility (AORs).

Organizing the Joint Force Headquarters

While each headquarters (HQ) organizes to accommodate the nature of the JFC's OA, mission, tasks, and preferences, all generally follow a traditional functional staff alignment (i.e., personnel, intelligence, operations, logistics, plans, and communications). HQ also have personal and special staff sections or elements, which perform specialized duties as prescribed by the JFC and handle special matters over which the JFC wishes to exercise personal control.

Organizing Operational Areas

Except for AORs, which are assigned in the Unified Command Plan (UCP), GCCs and other JFCs designate smaller OAs (e.g., joint operational area [JOA] and area of operations [AO]) on a temporary basis. OAs have physical dimensions comprised of some combination of air, land, maritime, and space domains. GCCs conduct operations in their assigned AORs. When warranted, the President, Secretary of Defense

Executive Summary

(SecDef), or GCCs may designate a theater of war and/or theater of operations for each operation. An AOR is an area established by the UCP that defines geographic responsibilities for a GCC. A theater of war is established primarily when there is a formal declaration of war or it is necessary to encompass more than one theater of operations (or a JOA and a separate theater of operations) within a single boundary for the purposes of C2, sustainment, protection, or mutual support. A theater of operations is an OA defined by the GCC for the conduct or support of specific military operations. For operations somewhat limited in scope and duration, or for specialized activities, the commander can establish the following OAs: JOA, joint special operations area, joint security area, amphibious objective area, and AO.

Joint Operations Across the Conflict Continuum

Introduction

Threats to US and allied interests throughout the world can sometimes only be countered by US forces able to respond to a wide variety of challenges along a conflict continuum that spans from peace to war.

Military Operations and Related Missions, Tasks, and Actions

Although the US military is organized, trained, and equipped for sustained, large-scale combat anywhere in the world, the capabilities to conduct these operations also enable a wide variety of other operations and activities.

In general, a military operation is a set of actions intended to accomplish a task or mission. Military operations are often categorized by their focus. Examples include stability activities; defense support to civil authorities; foreign humanitarian assistance (FHA); recovery; noncombatant evacuation operation (NEO); peace operations (PO); countering weapons of mass destruction; chemical, biological, radiological, and nuclear response; foreign internal defense (FID); counterdrug (CD) operations; combating terrorism; counterinsurgency (COIN); homeland defense (HD); and mass atrocities response.

The Range of Military Operations

The range of military operations is a fundamental construct that helps relate military activities

The range encompasses three primary categories: military engagement, security cooperation, and deterrence; crisis response and limited contingency operations; and large-scale combat operations. Military engagement, security cooperation, and deterrence activities develop local and regional

and operations in scope and purpose.

situational awareness, build networks and relationships with partners, shape the OE, keep day-to-day tensions between nations or groups below the threshold of armed conflict, and maintain US global influence. Many missions associated with crisis response and limited contingencies, such as defense support of civil authorities (DSCA) and FHA, may not require combat. But others, such as Operation RESTORE HOPE in Somalia, can be dangerous and may require combat operations to protect US forces. Large-scale combat often occurs in the form of major operations and campaigns that achieve national objectives or contribute to a larger, long-term effort (e.g., Operation ENDURING FREEDOM).

The Theater Campaign

Military operations, actions, and activities in a GCC's AOR, from security cooperation through large-scale combat, are conducted in the context of the GCC's ongoing theater campaign. The combatant commander's (CCDR's) theater campaign is the overarching framework that ensures all activities and operations within the theater are synchronized to achieve theater and national strategic objectives. A theater campaign plan operationalizes the GCC's strategy and approach to achieve these objectives within two to five years by organizing and aligning available resources.

A Joint Operation Model

Most individual joint operations share certain activities or actions in common. There are six general groups (shape, deter, seize initiative, dominate, stabilize, and enable civil authorities) of military activities that may typically occur in preparation for and during a single large-scale joint combat operation.

Phasing a Joint Operation

The six general groups of activity provide a convenient basis for thinking about a joint operation in notional phases. A phase is a definitive stage or period during a joint operation in which a large portion of the forces and capabilities are involved in similar or mutually supporting activities for a common purpose that often is represented by intermediate objectives. Phasing helps JFCs and staffs visualize, plan, and execute the entire operation and define requirements in terms of forces, resources, time, space, and purpose. Actual phases of an operation will vary (e.g., compressed, expanded, or omitted entirely) according to the nature of the operation and the JFC's decisions. Phases may be conducted sequentially, but some activities from a phase may begin in a previous phase and continue into subsequent phases.

The Balance of Offense, Defense, and Stability Activities

Most combat operations will require the commander to balance offensive, defensive, and stability activities. This is particularly evident in a campaign or major operation, where combat can occur during several phases and stability activities may occur throughout. Commanders strive to apply the many dimensions of military power simultaneously across the depth, breadth, and height of the OA. The challenge of balance and simultaneity affects all operations involving combat, particularly campaigns, due to their scope. Consequently, JFCs often concentrate in some areas or on specific functions, and require economy of force in others.

Linear and Nonlinear Operations

In linear operations, each commander directs and sustains combat power toward enemy forces in concert with adjacent units. Linearity refers primarily to the conduct of operations with identified forward lines of own troops. In linear operations, emphasis is placed on maintaining the position of friendly forces in relation to other friendly forces. From this relative positioning of forces, security is enhanced and massing of forces can be facilitated. **In nonlinear operations,** forces orient on objectives without geographic reference to adjacent forces. Nonlinear operations typically focus on creating specific effects on multiple decisive points. Nonlinear operations emphasize simultaneous operations along multiple LOOs from selected bases (ashore or afloat). Simultaneity overwhelms opposing C2 and allows the JFC to retain the initiative.

Executive Summary

Military Engagement, Security Cooperation, and Deterrence

Introduction

Military engagement, security cooperation, and deterrence activities provide the foundation of the CCDR's theater campaign. The goal is to prevent and deter conflict by keeping adversary activities within a desired state of cooperation and competition. **Military engagement** is the routine contact and interaction between individuals or elements of the Armed Forces of the United States and those of another nation's armed forces, or foreign and domestic civilian authorities or agencies, to build trust and confidence, share information, coordinate mutual activities, and maintain influence. **Security cooperation** involves all DOD interactions with foreign defense establishments to build defense relationships that promote specific US security interests, develop allied and friendly military capabilities for self-defense and multinational operations, and provide US forces with peacetime and contingency access to the host nation (HN). **Deterrence** prevents adversary action through the presentation of a credible threat of unacceptable counteraction and belief that the cost of the action outweighs the perceived benefits.

Typical Operations and Activities

Typical operations include: military engagement; emergency preparedness; arms control, nonproliferation, and disarmament; counterterrorism; support to CD operations; sanction enforcement; enforcement of exclusion zones; freedom of navigation and overflight; foreign assistance; security assistance; security force assistance; FID; humanitarian assistance programs; protection of shipping; show of force operations; support to insurgency; and COIN.

Other Considerations

Other considerations include: interagency, international, and nongovernmental organizations and HN coordination; information sharing; and cultural awareness.

Crisis Response and Limited Contingency Operations

Combatant commanders plan for various situations that require

When crises develop and the President directs, CCDRs respond. If the crisis revolves around

Executive Summary

military operations in response to natural disasters, terrorists, subversives, or other contingencies and crises as directed by appropriate authority. The level of complexity, duration, and resources depends on the circumstances.

Typical Operations

external threats to a regional partner, CCDRs employ joint forces to deter aggression and signal US commitment (e.g., deploying joint forces to train in Kuwait). If the crisis is caused by an internal conflict that threatens regional stability, US forces may intervene to restore or guarantee stability (e.g., Operation RESTORE DEMOCRACY, the 1994 intervention in Haiti). If the crisis is within US territory (e.g., natural or man-made disaster, deliberate attack), US joint forces will conduct DSCA and HD operations as directed by the President and SecDef.

NEOs are operations directed by the Department of State (DOS) or other appropriate authority, in conjunction with DOD, whereby noncombatants are evacuated from locations within foreign countries to safe havens designated by DOS when their lives are endangered by war, civil unrest, or natural disaster.

PO are multiagency and multinational operations involving all instruments of national power—including international humanitarian and reconstruction efforts and military missions—to contain conflict, restore the peace, and shape the environment to support reconciliation and rebuilding and facilitate the transition to legitimate governance.

FHA operations relieve or reduce human suffering, disease, hunger, or privation in countries outside the US. These operations are different from foreign assistance primarily because they occur on short notice as a contingency operation to provide aid in specific crises or similar events rather than as more deliberate foreign assistance programs to promote long-term stability.

Strikes are attacks conducted to damage or destroy an objective or a capability.

Raids are operations to temporarily seize an area, usually through forcible entry, in order to secure information, confuse an enemy, capture personnel or equipment, or destroy an objective or capability.

Executive Summary

HD is the protection of US sovereignty, territory, domestic population, and critical defense infrastructure against external threats and aggression or other threats as directed by the President. DOD is the federal agency with lead responsibility, supported by other agencies, to defend against external threats and aggression.

DSCA is support provided by US federal military forces, DOD civilians, DOD contract personnel, DOD component assets, DOD agencies, and National Guard forces (when SecDef, in coordination with the governors of the affected states, elects and requests to use those forces in Title 32, United States Code, status) in response to requests for assistance from civil authorities for domestic emergencies, law enforcement support, and other domestic activities, or from qualifying entities for special events. For DSCA operations, DOD supports and does not supplant civil authorities.

Large-Scale Combat Operations

Traditionally, campaigns are the most extensive joint operations, in terms of the amount of forces and other capabilities committed and duration of operations.

In the context of large-scale combat, a campaign is a series of related major operations aimed at achieving strategic and operational objectives within a given time and space. A major operation is a series of tactical actions, such as battles, engagements, and strikes, and is the primary building block of a campaign. Campaigns are joint in nature—functional and Service components of the joint force conduct supporting operations, not independent campaigns.

Combatant Command Planning

CCDRs document the full scope of their campaigns in the set of plans that includes the theater or functional campaign plan, and all of its GEF- and JSCP-directed plans, subordinate and supporting plans, posture or master plans, country plans (for the geographic CCMDs), operation plans of operations currently in execution, contingency plans, and crisis action plans.

Setting Conditions for Theater Operations

CCDRs and JFCs execute their campaigns and operations in pursuit of US national objectives and

Executive Summary

to shape the OE. In pursuit of national objectives, these campaigns and operations also seek to prevent, prepare for, or mitigate the impact of a crisis or contingency.

Considerations for Deterrence

The deter phase is characterized by preparatory actions that indicate resolve to commit resources and respond to the situation. Deterrence should be based on capability (having the means to influence behavior), credibility (maintaining a level of believability that the proposed actions may actually be employed), and communication (transmitting the intended message to the desired audience) to ensure greater effectiveness (effectiveness of deterrence must be viewed from the perspective of the agent/actor that is to be deterred). Considerations include: preparing the OA, isolating the enemy, flexible deterrent option and flexible response option, protection, space operations, geospatial intelligence support to operations, and physical environment.

Considerations for Seizing the Initiative

As operations commence, the JFC needs to exploit friendly advantages and capabilities to shock, demoralize, and disrupt the enemy immediately. The JFC seeks decisive advantage through the use of all available elements of combat power to seize and maintain the initiative, deny the enemy the opportunity to achieve its objectives, and generate in the enemy a sense of inevitable failure and defeat. Considerations for seizing the initiative include: force protection, unit integrity during deployment, entry operations, attack of enemy's COGs, full-spectrum superiority, C2 in littoral areas, SOF-conventional force integration, stability activities, protection, and prevention of friendly fire incidents.

Considerations for Dominance

JFCs conduct sustained combat operations when a swift victory is not possible. During sustained combat operations, JFCs simultaneously employ conventional forces and SOF throughout the OA. The JFC may designate one component or LOO to be the main effort, with other components providing support and other LOOs as supporting efforts. When conditions or plans change, the main

Executive Summary

effort might shift. Considerations for the dominance include: operating in littoral areas, attack on enemy's COGs, synchronization and/or integrating maneuver and interdiction, operations when weapons of mass destruction are employed or located, and stability activities.

Considerations for Stabilization

Operations in a stabilize phase typically begin with significant military involvement, to include some combat and the potential for longer-term occupation. Operations then move increasingly toward transitioning to an interim civilian authority and enabling civil authority as the threat wanes and civil infrastructures are reestablished. The JFC's mission accomplishment requires fully integrating US military operations with the efforts of interorganizational participants in a comprehensive approach to accomplish assigned and implied tasks. Considerations for the stabilization phase include: several LOOs may be initiated immediately (e.g., providing FHA, establishing security), forces and capabilities mix [need to realign], stability activities, measures to prevent complacency and be ready to counter activity that could bring harm to units or jeopardize the operation, restraint, perseverance, legitimacy, and operations security.

Considerations for Enabling Civil Authority

In this phase, the joint operation is assessed and enabling objectives are established for transitioning from large-scale combat operations to FID and security cooperation. The new government obtains legitimacy, and authority is transitioned from an interim civilian authority or transitional military authority to the new indigenous government. This situation may require a change in the joint operation as a result of an extension of the required stability activities in support of US diplomatic, HN, international organization, and/or NGO stabilization efforts. Considerations for enabling civil authority include: peace building, transfer to civil authority, and redeployment.

CONCLUSION

This publication is the keystone document of the joint operations series. It provides the doctrinal foundation and fundamental principles that guide the Armed Forces of the United States in all joint operations.

CHAPTER I
FUNDAMENTALS OF JOINT OPERATIONS

> *"The US military's purpose is to protect our Nation and win our wars. We do this through military operations to defend the homeland, build security globally, and project power and win decisively."*
>
> **The National Military Strategy of the United States of America, 2015**

1. Introduction

a. Joint Publication (JP) 3-0 is the keystone document in the joint operations series and is a companion to joint doctrine's capstone JP 1, *Doctrine for the Armed Forces of the United States*. It provides guidance to joint force commanders (JFCs) and their subordinates to plan, execute, and assess joint military operations. It also informs interagency and multinational partners, international organizations, nongovernmental organizations (NGOs), and other civilian decision makers of fundamental principles, precepts, and philosophies that guide the employment of the Armed Forces of the United States. This publication describes fundamental keystone constructs—such as unified action and joint functions—that apply regardless of the nature or circumstances of a specific joint operation. This publication provides context not only for the joint operations series, but also for other keystone doctrine publications that describe supporting functions and processes.

b. The primary way the Department of Defense (DOD) employs two or more Services (from at least two Military Departments) in a single operation is through joint operations. **Joint operations** are military actions conducted by joint forces and those Service forces employed in specified command relationships with each other, which of themselves do not establish joint forces. A **joint force** is one composed of significant elements, assigned or attached, of two or more Military Departments operating under a single JFC.

c. **Joint operations doctrine is built on warfighting philosophy and theory derived from experience.** Its foundation rests upon the principles of war and the associated fundamentals of joint warfare, described in JP 1, *Doctrine for the Armed Forces of the United States*. These principles and fundamentals apply to both traditional and irregular forms of warfare. Joint doctrine recognizes the utility of unity of command and the synergy created by the integration and synchronization of military operations in time, space, and purpose. Our leaders employ the Armed Forces of the United States—the military instrument of national power—in coordination with **diplomatic, informational, and economic** instruments to advance and defend US values and interests, achieve objectives consistent with national strategy, and conclude operations on terms favorable to the US.

Refer to JP 1, Doctrine for the Armed Forces of the United States, *for more information on traditional warfare and irregular warfare (IW) and the instruments of national power.*

Chapter I

d. **Joint Warfare is Team Warfare.** The Armed Forces of the United States—every military organization at all levels—are a team. The capacity of our Armed Forces to operate as a cohesive joint team is a key advantage in any operational environment (OE). Success depends on well-integrated command headquarters (HQ), supporting organizations, and forces that operate as a team. Integrating Service components' capabilities under a single JFC maximizes the effectiveness and efficiency of the force. However, a joint operation does not require that all forces participate merely because they are available; the JFC has the authority and responsibility to tailor forces to the mission.

e. **Principles of Joint Operations.** Joint doctrine recognizes **the nine principles of war.** Experience gained in a variety of IW situations has reinforced the value of three additional principles—**restraint, perseverance,** and **legitimacy.** Together, they comprise the **12 principles of joint operations** (see Figure I-1). See Appendix A, "Principles of Joint Operations."

f. **Common Operating Precepts.** In addition to the principles of joint operations, 10 **common operating precepts** underlie successful joint operations. Listed in Figure I-2, these precepts flow from broad challenges in the strategic environment to specific conditions, circumstances, and influences in a JFC's OE. The precepts can apply in all joint operations, although some may not be relevant activities such as military engagement.

2. Strategic Environment and National Security Challenges

a. The strategic environment consists of a variety of national, international, and global factors that affect the decisions of senior civilian and military leaders with respect to the employment of US instruments of national power in peace and periods of conflict. The strategic environment is uncertain, complex, and can change rapidly, requiring military leaders to maintain persistent military engagement with multinational partners. Although the basic character of war has not changed, the character of conflict has evolved. The military environment and the threats it presents are increasingly *transregional, multi-domain,* and *multi-functional* (TMM) in nature. By TMM we mean that the crises and contingencies joint forces face today cut across multiple combatant commands; cut across land, sea, air, space, and cyberspace; and involve conventional, special operations, ballistic

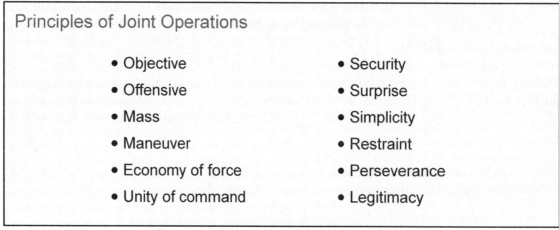

Figure I-1. Principles of Joint Operations

Fundamentals of Joint Operations

Common Operating Precepts

- Achieve and maintain unity of effort within the joint force and between the joint force and US Government, international, and other partners.
- Leverage the benefits of operating indirectly through partners when strategic and operational circumstances dictate or permit.
- Integrate joint capabilities to be complementary rather than merely additive.
- Focus on objectives whose achievement suggests the broadest and most enduring results.
- Ensure freedom of action.
- Avoid combining capabilities where doing so adds complexity without compensating advantage.
- Inform domestic audiences and shape the perceptions and attitudes of key foreign audiences as an explicit and continuous operational requirement.
- Maintain operational and organizational flexibility.
- Drive synergy to the lowest echelon at which it can be managed effectively.
- Plan for and manage operational transitions over time and space.

Figure I-2. Common Operating Precepts

missile, strike, cyber, and space capabilities. The strategic environment is fluid, with continually changing alliances, partnerships, and national and transnational threats that rapidly emerge, disaggregate, and reemerge. While it is impossible to predict precisely how challenges will emerge and what form they might take, we can expect that uncertainty, ambiguity, and surprise will persist. The commander's OE is influenced by the strategic environment.

   b. By acquiring advanced technologies, adversaries are changing the conditions of warfare that the US has become accustomed to in the past half century. Today's potential adversaries can increasingly synchronize, integrate, and direct lethal operations and other non-lethal elements of power with greater sophistication, and are less constrained by geographic, functional, legal, or phasing boundaries. Conflict is now, and will remain, inherently transregional as future potential adversaries' interests, influence, capabilities, and reach extend beyond single areas of operation. Significant and emerging challenges include, but are not limited to, traditional armed conflict, cyberspace attacks, terrorism involving weapons of mass destruction (WMD), adversary information operations (IO) campaigns, and proliferation of adversary antiaccess (A2) and area denial (AD) capabilities. A2 capabilities, usually long-range, prevent or inhibit an advancing force from entering an operational area (OA). If a force is able to overcome an enemy's A2 capabilities, additional AD capabilities can limit a force's freedom of action within an OA.

   c. These challenges are not specific to any single theater of operations and create problematic consequences for international security. Such an environment induces

Chapter I

instability; erodes the credibility of US national power; can necessitate escalation in US and allied responses; and weakens US alliances that promote trade, economic development, and diplomatic agreements. In the most challenging scenarios, the US may be unable to deploy, employ, and sustain forces the way it has in the recent past, i.e., build up combat power in an area, perform detailed rehearsals and integration activities, and then conduct operations when and where desired. JFCs should consider a wide range of options consistent with current realities of an OE not solely defined by state-on-state and force-on-force engagements. A JFC's OE, which encompasses all enemy, friendly, and neutral factors relevant to a specific joint operation, can include actions directed against a variety of state forces and non-state actors, to include insurgents, proxies, local warlords, criminals, and others. Civilians and organizations other than an enemy may also affect strategic outcomes. These actors may include the civilian population, host nation (HN) government, potential opposition leaders, international organizations, and NGOs.

d. Enemies who attack the US homeland and US interests are likely to use asymmetric tactics and techniques. They will avoid hard (well-secured and heavily defended) targets and attack vulnerable ones. Vulnerable targets may include US and partner nations' (PNs') lines of communications (LOCs), ports, airports, staging areas, civilian populations, critical infrastructure, information centers, economic centers, and military and police personnel and facilities. Advances in information technology increase the tempo, lethality, and depth of warfare. Developments in cyberspace can provide the US military, its allies, and PNs leverage to improve economic and physical security. However, this also provides adversaries increased access to open-source information and intelligence, Department of Defense information network (DODIN), critical infrastructure and key resources, and a limitless propaganda platform with global reach. Asymmetric attacks can be countered with well-planned joint operations synchronized with actions of interagency partners, international organizations, NGOs, multinational forces, and elements of the private sector. Achieving unity of effort with these partners requires coordination, cooperation, and a comprehensive approach to achieve common objectives.

Refer to JP 1, Doctrine for the Armed Forces of the United States, *and the* Defense Strategy Review (DSR) *for more information on the strategic security environment. Refer to JP 3-08,* Interorganizational Cooperation, *for more information on interorganizational coordination.*

3. Instruments of National Power and the Conflict Continuum

a. US **instruments of national power** are the national-level **means** our national leaders can apply in various **ways** to achieve strategic objectives **(ends).** Institutions that represent these instruments of national power are active continuously as the President directs along a conflict continuum that ranges from peace to war.

b. The ability of the US to advance its national interests depends on how the United States Government (USG) employs the instruments of national power to achieve national strategic objectives based on global security priorities. USG officials, with National Security Council (NSC) advice and presidential direction, coordinate the instruments of national power. The USG routinely uses the instruments of national power to advance

Fundamentals of Joint Operations

national interests. Interactions between the various instruments of national power can enhance results as US culture, industry, science and technology, academic institutions, geography, and national will combine to deliver synergistic benefit.

c. The ultimate purpose of the US Armed Forces is to fight and win the nation's wars. Although much of DOD's focus is on war and war preparation, opportunities also exist to prevent or mitigate the severity of conflict, legitimize US positions, reward PNs, provide expertise to multinational operations, and enhance the positive perception of the US. US national leaders can use military capabilities in a wide variety of activities, tasks, missions, and operations that vary in purpose, scale, risk, and combat intensity along the **conflict continuum.** The military's role increases relative to the other instruments as the need to compel an adversary through force increases. The potential **range of military activities and operations** extends from military engagement, security cooperation, and deterrence in times of relative peace up through major operations and campaigns that typically involve large-scale combat. For more information on the range of military activities and operations across the conflict continuum, see Chapter V, "Joint Operations Across the Conflict Continuum," paragraph 2, "The Range of Military Operations."

d. Acting alone in the strategic environment, the USG cannot resolve all crises or achieve all national objectives with just US resources. Under an umbrella of security cooperation, DOD supports USG strategic objectives by developing security relationships, building partner capacity and capability, and assuring access with selected PNs that enable them to act alongside, in support of, or in lieu of US forces around the globe. These strategic initiatives help advance national security objectives, promote stability, prevent conflicts, and reduce the risk of employing US military forces in a conflict. Security cooperation activities comprise an essential element of a geographic combatant commander's (GCC's) theater campaign plan (TCP).

Refer to JP 1, Doctrine for the Armed Forces of the United States, *for more information on the instruments of national power. Refer to JP 3-20,* Security Cooperation, *for more information about a GCC's role in security cooperation. Refer to Chapter V, "Joint Operations Across the Conflict Continuum," for more information on the conflict continuum and range of military operations.*

4. Strategic Direction

a. **National strategic direction** is governed by the Constitution, federal law, USG policy, internationally recognized law, and the national interest as represented by national security policy. This direction **provides strategic context** for the employment of the instruments of national power and defines the strategic purpose that guides employment of the military as part of a global strategy. Strategic direction is typically published in key documents, generally referred to as strategic guidance, but it may be communicated through any means available. Strategic direction may change rapidly in response to changes in the global environment, whereas strategic guidance documents are typically updated cyclically and may not reflect the most current strategic direction.

Chapter I

(1) In general, the President frames the strategic context by defining national interests and goals in documents such as the National Security Strategy (NSS), Presidential policy directives, executive orders, and other national strategic documents, in conjunction with the NSC and Homeland Security Council.

(2) DOD derives its strategic-level documents from guidance in the NSS. The documents outline how DOD will support NSS objectives and provide a framework for other DOD policy and planning guidance, such as the Guidance for Employment of the Force (GEF), Defense Planning Guidance, Global Force Management Guidance, and the Joint Strategic Capabilities Plan (JSCP).

(3) The President approves the contingency planning guidance contained in the GEF, which is developed by the Office of the Secretary of Defense. The GEF provides policy guidance and priorities to the Chairman of the Joint Chiefs of Staff (CJCS) and combatant commanders (CCDRs) for global force management and the preparation and review of campaign and contingency plans. The CJCS translates guidance from the GEF and publishes the JSCP, which implements campaign, contingency, and posture planning guidance reflected in the GEF. The President also signs the Unified Command Plan (UCP), which is developed by the Office of the Secretary of Defense and the Joint Staff in coordination with the NSC. The UCP establishes combatant command (CCMD) missions, responsibilities, and areas of responsibility (AORs).

b. From this broad strategic guidance, more specific national, functional, and theater-strategic and supporting objectives help focus and refine the context and guide the military's joint planning and execution related to these objectives or a specific crisis. Integrated planning, coordination, and guidance among the Joint Staff, CCMD staffs, Service chiefs, and USG departments and agencies translate strategic priorities into clear planning guidance, tailored force packages, operational-level objectives, joint operation plans (OPLANs), and logistical support for the joint force to accomplish its mission.

For more information on national strategic direction, refer to Presidential Policy Directive-1, Organization of the National Security System; *Chairman of the Joint Chiefs of Staff Instruction (CJCSI) 5715.01,* Joint Staff Participation in Interagency Affairs; *JP 1,* Doctrine for the Armed Forces of the United States; *and JP 5-0,* Joint Planning.

c. **Commander's communication synchronization (CCS)** is a process that helps implement strategic-level guidance by coordinating, synchronizing, and ensuring the integrity and consistency of strategic- to tactical-level narratives, themes, messages, images, and actions throughout a joint operation across all relevant communication activities. JFCs, their component commanders, and staffs coordinate and adjust CCS plans, programs, products, and actions with the other interorganizational participants employed throughout the OA, such as the various chiefs of mission relevant to the joint operation. Effective CCS focuses processes and efforts to understand and communicate with key audiences and create, strengthen, or preserve conditions favorable to advance USG interests, policies, and objectives. See paragraph 2.j., "Commander's Communication Synchronization," in Chapter III, "Joint Functions," for more information.

Fundamentals of Joint Operations

Refer to JP 1, Doctrine for the Armed Forces of the United States; *JP 3-61,* Public Affairs; *JP 5-0,* Joint Planning; *and Joint Doctrine Note (JDN) 2-13,* Commander's Communication Synchronization, *for more information on CCS.*

d. **The CCDR's Strategic Role**

(1) Based on guidance from the President and the Secretary of Defense (SecDef), GCCs and functional combatant commanders (FCCs) translate national security policy, strategy, and available military forces into theater and functional strategies to achieve national and theater strategic objectives. CCMD strategies are broad statements of the GCC's long-term vision for the AOR and the FCC's long-term vision for the global employment of functional capabilities guided by and prepared in the context of the SecDef's priorities outlined in the GEF and the CJCS's objectives articulated in the National Military Strategy (NMS). A prerequisite to preparing the theater strategy is development of a strategic estimate. It contains factors and trends that influence the CCMD's strategic environment and inform the ends, ways, means, and risk involved in pursuit of GEF-directed objectives.

(2) Using their strategic estimates and theater or functional strategies, GCCs and FCCs develop TCPs and functional campaign plans (FCPs) respectively, consistent with guidance in the UCP, GEF, and JSCP, as well as in accordance with (IAW) planning architecture described in the Adaptive Planning and Execution (APEX) enterprise. In some cases, a CCDR may be required to develop a global campaign plan. FCCs develop operational support plans based on guidance in the UCP and their priorities and objectives in the GEF. FCCs may be responsible for developing functional-related global or subordinate campaign plans or both. As required, both GCCs and FCCs develop contingency plans, which are branch plans to the overarching TCP or FCP.

(3) In joint operations, the supported CCDR will often have a role in achieving more than one national strategic objective. Some national strategic objectives will be the primary responsibility of the supported CCDR. Others may require a more balanced use of many or all instruments of national power, with the CCDR in support of another CCDR or other agencies. Supporting CCDRs coordinate and synchronize their supporting plans with the supported commander's plan. CCDRs provide planning guidance; assign missions and tasks; organize forces and resources; designate objectives; may establish operational limitations, such as rules of engagement (ROE), constraints, and restraints; and implement policies and the concept of operations (CONOPS) to be integrated into plans and operation orders (OPORDs). In applying military power, CCDRs use the capabilities of assigned, attached, and supporting military forces. They also integrate other instruments of national power and the capabilities of multinational partners to gain and maintain strategic advantage. Supporting and supported CCDRs coordinate with each other across geographic and functional boundaries to facilitate mission accomplishment within the capabilities of assigned, attached, and supporting military forces.

Refer to JP 1, Doctrine for the Armed Forces of the United States, *for more information on the strategic environment, strategic guidance, strategy and estimates, and the role of*

Chapter I

CCDRs. For more information on APEX and joint planning, refer to CJCS Guide 3130, Adaptive Planning and Execution Overview and Policy Framework.

(4) **Termination**

(a) **Terminating joint operations** and transforming conflict into lasting stability is an aspect of the CCDR's functional or theater strategy that links to achievement of national objectives. Successful military operations can create conditions so that the other instruments of national power can achieve national security objectives. Based on the President's strategic objectives, the supported CCDR develops and proposes **termination criteria**—the specified conditions approved by the President or SecDef that must be met before a named operation or campaign can be concluded. These termination criteria help define the desired **military end state,** which normally represents a point in time or a set of conditions beyond which the President does not require the military instrument of national power as the primary means to achieve remaining national objectives. This period or set of circumstances usually signals a transition from military to civilian lead of subsequent activities.

(b) Creating the conditions for lasting stability extends beyond the large-scale combat that dominates over the enemy's will to resist. Often, overmatching the enemy's capabilities at critical times and places does not lead to the resolution of the drivers of conflict, nor initiate the transformation to eliminate the root causes of conflict. The Armed Forces often remain engaged in a supporting role to assist in reducing the means and motives for violence, focusing on the sources of instability, developing opportunities to promote stability. Strategic- and operational-level commanders plan for the termination of operations to enable civil authority as part of the initial planning process. Understanding transitions is key to operational design and planning.

For more information on end state and termination, refer to JP 5-0, Joint Planning.

5. Unified Action

a. **General**

(1) Whereas the term **joint operation** focuses on the integrated actions of the Armed Forces of the United States, the term **unified action** has a broader connotation. Unified action refers to the synchronization, coordination, and integration of the activities of governmental and nongovernmental entities to **achieve unity of effort.** Failure to achieve unity of effort can cost lives, create conditions that enhance instability, and jeopardize mission accomplishment.

COMMON OPERATING PRECEPT

Achieve and maintain unity of effort between the joint force and interorganizational participants.

I-8 JP 3-0

Fundamentals of Joint Operations

(2) Unified action is based on national strategic direction, which is governed by the Constitution, federal law, and USG policy. Unified action is a comprehensive approach that focuses on coordination and cooperation of the US military and other interorganizational participants toward common objectives, even if the participants are not necessarily part of the same command or organization.

(3) Enabled by the principle of **unity of command,** military leaders understand the effective mechanisms to achieve **military** unity of effort. The goal of unified action is to achieve a similar unity of effort between participants, as Figure I-3 shows. This publication uses the term **interorganizational participants** to refer collectively to USG departments and agencies (i.e., interagency partners); state, territorial, local, and tribal agencies; foreign military forces and government agencies (i.e., multinational partners); NGOs; and the private sector. This aligns with the term and definition of interorganizational cooperation established in JP 3-08, *Interorganizational Cooperation*.

(4) The US Department of State (DOS) has a complementary approach, which defines unity of effort as a cooperative concept that refers to coordination and communication among USG organizations toward the same common goals for success.

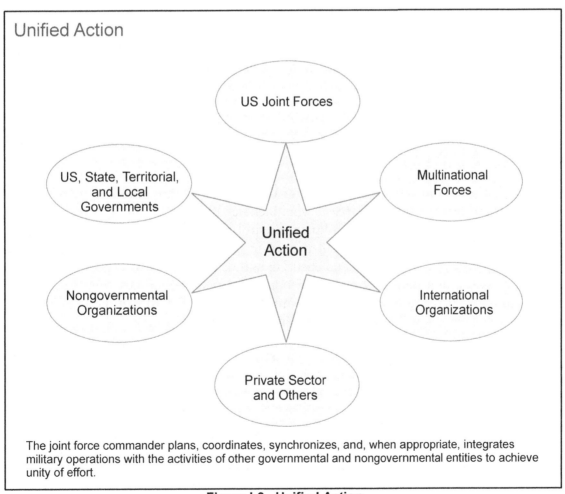

Figure I-3. Unified Action

Chapter I

The basis is the necessity of each agency's efforts to be in harmony with the short- and long-range goals of the mission.

b. **The JFC's Role.** JFCs are challenged to achieve and maintain operational coherence given the requirement to operate in conjunction with interorganizational partners. CCDRs play a pivotal role in unifying joint force actions, since all of the elements and actions that compose unified action normally are present at their level. However, subordinate JFCs also integrate and synchronize their operations directly with the operations of other military forces and the activities of nonmilitary organizations in the operational area to promote unified action.

c. **Multinational Participation in Unified Action**

(1) **General.** Joint forces must be prepared to plan and execute operations with forces from PNs within the framework of an alliance or coalition under US or other-than-US leadership. US military leaders often are expected to play a central leadership role regardless of the US Armed Forces' predominance, capability, or capacity. Commanders should expect the military leaders of contributing member nations to emphasize common objectives as well as to expect mutual support and respect. Although individual nations may place greater emphasis on some objectives than on others, the key is to find commonality within the objectives to promote synchronized progress to achieving the objectives. Cultivation and maintenance of personal relationships among counterparts enable success. Language and communication differences, cultural diversity, historical animosities, and the varying capabilities of allies and multinational partners are factors that complicate the integration and synchronization of activities during multinational operations. Likewise, differing national obligations derived from international treaties, agreements, and national legislation complicate multinational operations. Regardless of whether other members participate in their treaty or agreement obligations, US forces will remain bound by US treaties and agreements.

(2) **Command and Control (C2) of US Forces. By law, the President retains command authority over US forces.** This includes the authority and responsibility to effectively plan for, organize, coordinate, control, employ, and protect these forces. Nevertheless, the President may deem it prudent or advantageous (for reasons such as maximizing military effectiveness and ensuring unified action) to place specific US forces under the control of a foreign commander to achieve specified military objectives. Even when operating under the operational control (OPCON) of a foreign commander, US Armed Forces remain in the chain of command of US military authorities.

(3) **C2 Structures.** Alliances typically develop C2 structures, systems, and procedures, with the predominant contributing nation providing the allied force commander. Staffs are integrated, and subordinate commands often are led by senior representatives from member nations. Shared doctrine, standardization agreements, close military cooperation, training, and robust diplomatic relations should characterize alliances. Coalitions are less standardized and may adopt a parallel or lead-nation C2 structure or a combination of the two. In a **parallel command construct**, nations retain control of their deployed forces and operate under their own doctrine and procedures, and

in a **lead nation command** construct, the PN providing the preponderance of forces and resources typically provides the commander of the coalition force. These command structures can also exist simultaneously within a coalition.

For more information on unified action with respect to multinational participation, refer to JP 1, Doctrine for the Armed Forces of the United States. *For more information on all aspects of multinational operations, refer to JP 3-16,* Multinational Operations. *For more information on multinational logistics, refer to JP 4-08,* Logistics in Support of Multinational Operations. *For North Atlantic Treaty Organization (NATO)-specific doctrine ratified by the US, see Allied Joint Publication (AJP)-01,* Allied Joint Doctrine, *and AJP-3,* Allied Joint Doctrine for the Conduct of Operations.

d. **Interorganizational Coordination in Unified Action**

(1) **General.** Previous paragraphs summarized US interaction with multinational partners. CCDRs and subordinate JFCs often interact with a variety of other interorganizational participants. This interaction varies according to the nature of participant (capability, capacity, objectives, etc.) and type of operation. JFCs and planners consider the potential contributions of other agencies and determine which can best contribute to achieving specific objectives. Often, other interagency partners, primarily DOS, can facilitate a JFC's coordination with multinational and HN agencies, NGOs, and the private sector. DOD may support other agencies during operations; however, under US law, US military forces will remain under the DOD command structure. Federal lead agency responsibility may be prescribed by law or regulation, Presidential directive, policy, or agreement among or between agencies. Even then, because of its resources and well-established planning methods, the joint force will likely provide significant support to the lead agency.

(2) **Civil-Military Integration**

(a) Military operations require civil-military integration. The degree of integration depends on the mission, objectives, organizations, governments, and people involved. Presidential directives guide participation by all USG departments and agencies. Military leaders work with the other USG partners to promote unified action. Differences in policies, procedure, decision-making processes, terminology, organizational cultures, as well as the nature and extent of resourcing across the various USG departments and agencies, will often complicate and may initially create some conflict challenges to successful civil-military integration.

(b) **Integration, cooperation, and coordination between military forces and interorganizational participants are much less structured than military C2.** Some

---

COMMON OPERATING PRECEPT

Leverage the benefits of operating indirectly through partners when strategic and operational circumstances dictate or permit.

Chapter I

organizations may have policies that conflict with those of the USG, particularly those of the US military. Formal agreements, robust liaison, and information sharing through disciplined interorganizational coordination should facilitate common understanding, informed decision making, and unity of effort. Information sharing with NGOs and the private sector may be more restrictive, but options such as the CCMD-level joint interagency coordination group (JIACG) and operational-level civil-military operations center (CMOC) are available to the commanders to facilitate interorganizational coordination and information sharing. DOD, in collaboration with federal, state, local, territorial, and tribal governments, uses the structures and procedures provided by the National Response Framework (NRF) and the National Incident Management System to prepare for, plan, coordinate, and respond to disasters within the US and its territories. Similar structures and processes, incorporating the capabilities and interests of foreign partners, can be incorporated into disaster-response and civil-military operations (CMO) in connection with US operations in foreign countries.

For more information on interorganizational coordination, refer to JP 3-08, Interorganizational Cooperation. *For more information on CMO and the CMOC, refer to JP 3-57,* Civil-Military Operations. *For more information on DOD's coordination and involvement with other government agencies within the context of homeland security and role within the NRF refer to JP 3-28,* Defense Support of Civil Authorities.

e. The US has neither the capacity nor responsibility to directly lead the response to every crisis. US military operations abroad invite diplomatic repercussions from enemies, adversaries, and even allies with whom our objectives do not precisely align. **In some circumstances, friendly surrogates assisted by US military support may be able to conduct operations and achieve mutually agreeable objectives when the direct employment of US forces would be objectionable or infeasible.** In other instances, such as counterinsurgency (COIN), success depends on the indigenous government demonstrating its own sovereignty; the overt exercise of force by the US military may ultimately be counterproductive. JFCs may increasingly find it advantageous or necessary to pursue objectives by enabling and supporting one or more interorganizational partners.

6. Levels of Warfare

a. **General.** Three levels of warfare—strategic, operational, and tactical—model the relationship between national objectives and tactical actions. There are no fixed limits or boundaries between these levels, but they help commanders visualize a logical arrangement of operations, allocate resources, and assign tasks to appropriate commands. Echelon of command, size of units, types of equipment, and types and location of forces or components may often be associated with a particular level, but the strategic, operational, or tactical purpose of their employment depends on the nature of their task, mission, or objective. For example, intelligence and communications satellites, previously considered principally strategic assets, are also significant resources for tactical operations. Likewise, tactical actions can cause both intended and unintended strategic consequences, particularly in today's environment of pervasive and immediate global communications and networked threats. Figure I-4 depicts the role of operational art in linking tactical- and operational-

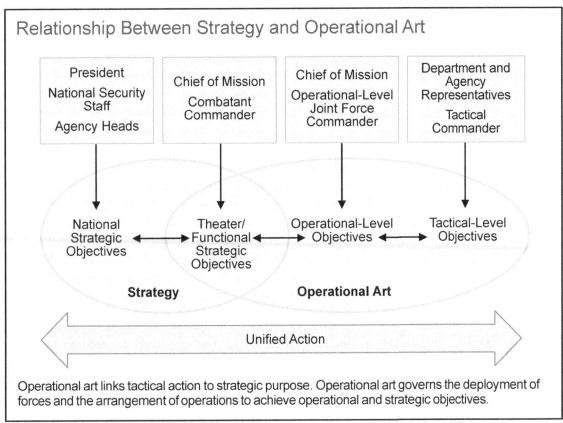

Figure I-4. Relationship Between Strategy and Operational Art

level actions and objectives to strategic objectives, and highlights leaders who typically focus at these levels.

b. **Strategic Level of Warfare.** In the context of national interests, strategy develops an idea or set of ideas of the ways to employ the instruments of national power in a synchronized and integrated fashion to achieve national, multinational, and theater objectives. Through development of strategy (e.g., the NSS, DSR, and NMS), a nation's leader, often with other nations' leaders, determines national or multinational strategic objectives with specific guidance to shape and allocate national resources to achieve these objectives. The President, aided by the NSC, establishes policy and national strategic objectives. SecDef translates these objectives into strategic military objectives that facilitate theater strategic planning. CCDRs usually participate in strategic discussions with the President and SecDef through the CJCS. CCDRs also participate in strategic discussions with allies and multinational members. **Thus, the CCDR's strategy relates to both US national strategy and operational-level activities within the theater.** Military strategy, derived from national policy and strategy and informed by doctrine, provides a framework for conducting operations.

c. **Operational Level**

(1) The operational level of warfare links the tactical employment of forces to national strategic objectives. **The focus at this level is on the planning and execution of operations using operational art:** the cognitive approach by commanders and staffs—

Chapter I

supported by their skill, knowledge, and experience—to plan and execute (when required) strategies, campaigns, and operations to organize and employ military capabilities by integrating ends, ways, and available means. JFCs and component commanders use operational art to determine how, when, where, and for what purpose military forces will be employed, to influence the adversary's disposition before combat, to deter adversaries from supporting enemy activities, and to assure our multinational partners to achieve operational and strategic objectives.

(2) Many factors affect relationships among leaders at these levels. Service and functional component commanders of a joint force do not plan the actions of their forces in a vacuum; they and their staffs collaborate with the operational-level JFC to plan the joint operation. This collaboration facilitates the components' planning and execution. Likewise, the operational-level JFC and staff typically collaborate with the CCDR to frame theater strategic objectives, as well as tasks the CCDR will eventually assign to the subordinate joint force.

d. **Tactical Level.** Tactics is the employment, ordered arrangement, and directed actions of forces in relation to each other. Joint doctrine focuses this term on planning and executing battles, engagements, and activities at the tactical level to achieve military objectives assigned to tactical units or task forces (TFs). An engagement can include a wide variety of noncombat tasks and activities and combat between opposing forces normally of short duration. A battle consists of a set of related engagements. Battles typically last longer than engagements, involve larger forces, and have greater potential to affect the course of a campaign.

7. Characterizing Military Operations and Activities

a. The US employs military capabilities in support of national security objectives in a variety of military operations and activities. The purpose of military action may be specified in a mission statement or implied from an order. Operations and activities are characterized as "joint" when they are conducted by a force composed of significant elements, assigned or attached, of two or more Military Departments operating under a single JFC.

b. Distinct military operations and activities may occur simultaneously with or independently of others even within the same OA. For example, a noncombatant evacuation operation (NEO) may be in the same OA where US forces are conducting COIN operations. Additionally, each may have different root causes and objectives. Chapter V, "Joint Operations Across the Conflict Continuum," discusses the simultaneous nature of theater operations and activities, and describes various joint operations and considerations in the context of the three broad areas of the range of military operations.

CHAPTER II
THE ART OF JOINT COMMAND

> *"When all is said and done, it is really the commander's coup d'oeil, his ability to see things simply, to identify the whole business of war completely with himself, that is the essence of good generalship."*
>
> **Carl von Clausewitz**
> *On War*

1. Introduction

a. **Command** is the authority that a commander in the armed forces lawfully exercises over subordinates by virtue of rank or assignment. Accompanying this authority is the responsibility to effectively organize, direct, coordinate, and control military forces to accomplish assigned missions. Command includes responsibility for health, welfare, morale, and discipline of assigned personnel.

b. While command authority stems from orders and other directives, **the art of command resides in the commander's ability to use leadership to maximize performance.** The combination of courage, ethical leadership, judgment, intuition, situational awareness, and the capacity to consider contrary views, helps commanders make insightful decisions in complex situations. These attributes can be gained over time through training, education, and experience. Joint training and joint doctrine are designed to enable the conscious and skillful exercise of command authority through visualization, decision making, and leadership. Effective commanders combine judgment and visualization with information to determine whether a decision is required, when to decide, and what to decide with sufficient speed to maintain the initiative. Information management (IM), situational awareness, and a sound battle rhythm facilitate decision making.

2. Commander-Centric Leadership

a. A commander's perspective of challenges in the OE is broad and comprehensive due to the interaction with USG civilian leaders; senior, peer, subordinate, and supporting commanders; and interorganizational partners. Clear commander's guidance and intent, enriched by the commander's experience and intuition, enable joint forces to achieve objectives. Employing the "art of war," which has been the commander's central historical command role, remains critical regardless of technological and informational improvements in control—the "science of war."

b. The C2 function is commander-centric and network-enabled to facilitate initiative and decision making at the lowest appropriate level. Although joint forces have grown accustomed to communicating freely without fear of jamming or interception, US enemies and adversaries are likely to use technological advances in cyberspace and vulnerabilities in the electromagnetic spectrum (EMS) to conduct cyberspace or EMS attacks. Commanders should be prepared to operate in an environment degraded by

Chapter II

electromagnetic interference. This is especially true at the lower echelons. If a commander loses reliable communications, **mission command**—a key component of the C2 function described in Chapter III, "Joint Functions"—enables military operations through decentralized execution based on mission-type orders. Mission command is built on subordinate leaders at all echelons who exercise disciplined initiative and act aggressively and independently to accomplish the mission. Mission-type orders focus on the purpose of the operation rather than details of how to perform assigned tasks. Commanders delegate decisions to subordinates wherever possible, which minimizes detailed control and empowers subordinates' initiative to make decisions based on the commander's guidance rather than constant communications. Subordinates' understanding of the **commander's intent** at every level of command is essential to mission command. See paragraph 5.e., "Key Planning Elements," for a discussion of commander's intent as a planning element.

c. Commanders should interact with other leaders to build personal relationships and develop trust and confidence. Developing these associations is a conscious, collaborative act. Commanders build trust through words and actions, and continue to reinforce it not only during operations, but also during training, education, and practice. Trust and confidence are essential to synergy and harmony, both within the joint force and with our interagency and multinational partners and other interorganizational stakeholders. Commanders may also interact with other political, societal, and economic leaders and other influential people who may influence joint operations. This interaction supports mission accomplishment and CCS themes and messages. The JFC emphasizes the importance of key leader engagement (KLE) to subordinate commanders and encourages them to extend the process to lower levels, based on mission requirements.

d. Commanders should provide subordinate commands sufficient time to plan, particularly in a time-sensitive crisis situation. They do so by issuing a warning order to subordinates at the earliest opportunity and by collaborating with other commanders, agency leaders, and multinational partners to develop a clear understanding of the commander's mission, intent, guidance, and priorities. Commanders resolve issues that are beyond the staff's authority. Examples include highly classified, limited-access planning for sensitive operations and allowing multinational partners' planners restricted access to US classified information systems.

e. Commanders collaborate with their seniors and peers to resolve differences of interpretation of higher-level objectives and the ways and means to accomplish these objectives. Commanders generally expect that their higher HQ has accurately described the OE, framed the problem, and devised a sound approach to achieve the best solution. Strategic guidance, however, can be vague, and the commander must interpret and clarify it for the staff. While national leaders and CCDRs may have a broader perspective of the problem, subordinate JFCs and their component commanders often have a better perspective of the situation at the operational level. Both perspectives are essential to a sound solution. During a commander's decision cycle, subordinate commanders should aggressively share their perspective with senior leaders to resolve issues at the earliest opportunity.

The Art of Joint Command

> **COMMON OPERATING PRECEPT**
>
> **Integrate joint capabilities to be complementary rather than merely additive.**

f. An essential skill of a JFC is the ability to assign missions and tasks that integrate the components' capabilities consistent with the JFC's envisioned CONOPS. Each component's mission should complement the others'. This enables each component to enhance the capabilities and limit the vulnerabilities of the others. Achieving this synergy requires more than just understanding the capabilities and limitations of each component. The JFC should also visualize operations holistically, identify the preconditions that enable each component to optimize its own contribution, and then determine how the other components might help to produce them. The JFC should compare alternative component missions and mixes solely from the perspective of combined effectiveness, unhampered by Service parochialism. This approach also requires mutual trust among commanders that the missions assigned to components will be consistent with their capabilities and limitations, those capabilities will not be risked for insufficient overall return, and components will execute their assignments.

g. Successful leaders encourage the exchange of information and ideas throughout their staffs to ensure decisions are based on the best understanding of the situation and available options. Such exchanges promote critical reviews of assumptions; facilitate consideration of all aspects of the situation, including cultural issues; stimulate broad consideration of military and nonmilitary alternatives; and emphasize efforts to minimize organizational and human sources of error and bias.

h. The JFC leads using operational art and operational design, joint planning, rigorous assessment of progress, and timely decision making.

3. Operational Art

a. **Operational art is the cognitive approach by commanders and staffs–supported by their skill, knowledge, experience, creativity, and judgment–to develop strategies, campaigns, and operations to organize and employ military forces by integrating ends, ways, and means.** It is a thought process to mitigate the ambiguity and uncertainty of a complex OE and develop insight into the problems at hand. Operational art also promotes unified action by enabling JFCs and staffs to consider the capabilities, actions, goals, priorities, and operating processes of interagency partners and other interorganizational participants, when they determine objectives, establish priorities, and assign tasks to subordinate forces. It facilitates the coordination, synchronization, and, where appropriate, the integration of military operations with activities of other participants, thereby promoting unity of effort.

b. The foundation of operational art encompasses broad vision; the ability to anticipate; and the skill to plan, prepare, execute, and assess. It helps commanders and their staffs organize their thoughts and envision the conditions necessary to accomplish the

Chapter II

mission and reach the desired military end state in support of national objectives. Without operational art, campaigns and operations could be sets of disconnected events. Operational art informs the deployment of forces and the arrangement of operations to achieve military operational and strategic objectives.

c. The commander is the central figure in operational art, not only due to education and experience, but also because the commander's judgment and decisions guide the staff throughout joint planning and execution. Commanders leverage their knowledge, experience, judgment, and intuition to focus effort and achieve success. Operational art helps broaden perspectives to deepen understanding and enable visualization. Commanders compare similarities of the existing situation with their own experiences or history to distinguish unique features and then tailor innovative and adaptive solutions to each situation.

d. The commander's ability to think creatively enhances the ability to employ operational art in order to answer the following questions:

(1) What are the **objectives and desired military end state?** (Ends)

(2) What **sequence of actions** is most likely to achieve those objectives and military end state? (Ways)

(3) What **resources** are required to accomplish that sequence of actions? (Means)

(4) What is the likely **chance of failure or unacceptable results** in performing that sequence of actions? (Risk)

e. Operational art encompasses **operational design—the conception and construction of the framework that underpins a joint operation or campaign plan and its subsequent execution.** Together, operational art and operational design strengthen the relationship between strategic objectives and the tactics employed to achieve them.

4. Operational Design

a. **Operational design** is the conception and construction of the framework that underpins a campaign or major operation plan and its subsequent execution. It extends operational art's vision with a creative process to help commanders and planners answer the ends-ways-means-risk questions. Commanders and staffs can use operational design when planning any joint operation.

b. Operational design supports operational art with a methodology designed to enhance understanding the situation and the problem. The methodology helps the JFC and staff identify broad solutions for mission accomplishment. Elements of operational design—such as **objective, center of gravity (COG), line of operation (LOO), line of effort (LOE), and termination**—are tools that help the JFC and the staff visualize and describe the broad operational approach to achieve objectives and accomplish the mission. These operational design elements are useful throughout the joint planning process (JPP).

Using these elements helps the commander and staff analyze the questions posed in paragraph 3.d. For example, properly framing strategic- and operational-level objectives is essential to mission accomplishment. Identifying critical objectives will depend more on judgment than on calculation, because framing objectives to achieve broad and enduring results is more art than science.

c. Operational design works best when commanders encourage discourse and leverage dialogue and collaboration to identify and solve complex, ill-defined problems. To that end, the commander should empower organizational learning and develop methods to determine whether the operational approach should be modified during the course of an operation. This requires continuous assessment and reflection that challenge understanding of the existing problem and the relevance of actions addressing that problem.

d. Commanders and their staffs blend operational art, operational design, and JPP to produce plans and orders that drive joint operations. Effective operational design results in more efficient detailed planning and increases the chances of mission accomplishment.

Refer to JP 5-0, Joint Planning, for the details of operational design.

5. Joint Planning

a. Planning translates guidance into plans or orders to achieve a desired objective or attain an end state. The joint planning and execution community begins planning when a potential or actual event is recognized that may require a military response. Objectives provide a unifying purpose around which to focus actions and resources. JPP aligns military activities and resources to achieve national objectives and enables leaders to examine cost-benefit relationships, risks, and trade-offs to determine a preferred course of action (COA) to achieve that objective or attain an end state. Joint planning occurs within the Joint Operation Planning and Execution System and the APEX enterprise, which encompasses department-level joint planning policies, processes, procedures, and reporting structures.

b. Joint planning consists of planning activities that help CCDRs and their subordinate commanders transform national objectives into actions that mobilize, deploy, employ, sustain, redeploy, and demobilize joint forces. It ties the employment of the Armed Forces to the achievement of national objectives during peacetime and war.

c. Based on understanding gained through the application of operational design, more detailed planning takes place within the steps of JPP. JPP is an orderly, analytical process, which consists of a set of logical steps to analyze a mission; develop, analyze, and compare alternative COAs; select the best COA; and produce a plan or order.

d. **JPP underpins planning at all levels and for missions across the range of military operations.** It applies to both supported and supporting JFCs and to component and subordinate commands when they participate in joint planning. Together with operational design, JPP facilitates interaction between the commander, staff, and their HQ throughout planning. JPP helps commanders and their staffs organize their planning

Chapter II

activities, share a common understanding of the mission and commander's intent, and develop effective plans and orders. Figure II-1 shows the primary steps of JPP.

e. **Key Planning Elements.** Commanders participate in planning to the greatest extent possible from early operational design through approval of the plan or order. **Regardless of the commander's level of involvement, certain key planning elements require the commander's participation and decisions.** These include the operational approach, mission statement, commander's planning guidance, commander's intent, commander's critical information requirements (CCIRs), and CONOPS.

(1) **Operational Approach.** The operational approach is a commander's initial description, to help guide further planning, of the broad actions the force must take to achieve objectives and accomplish the mission. It is the commander's visualization of how the operation should transform current conditions into the desired conditions—the way the commander wants the OE to look at the conclusion of operations. The operational approach is based largely on an understanding of the OE and the problem facing the JFC. Once the JFC approves the approach, it provides the basis to begin, continue, or complete detailed planning. The JFC and staff should continually review, update, and modify the approach as the OE, objectives, or problem change.

(2) **Mission Statement.** The joint force's mission is what the joint force must accomplish. It is described in the mission statement, which is a sentence or short paragraph that describes the organization's essential task (or set of tasks) and purpose—a clear statement of the action to be taken and the reason for doing so. The mission statement— approved by the commander—contains the elements of who, what, when, where, and why of the operation. The eventual CONOPS will specify how the joint force will accomplish the mission. The mission statement forms the basis for planning and is included in the

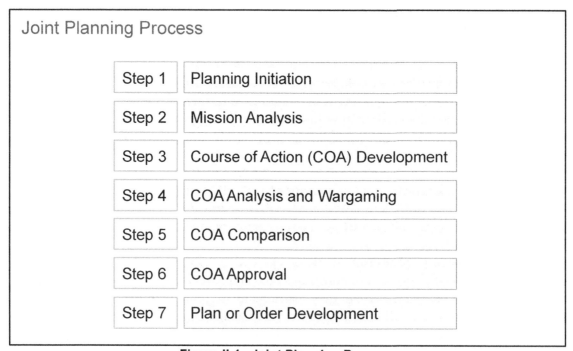

Figure II-1. Joint Planning Process

commander's planning guidance, the planning directive, staff estimates, commander's estimate, and the CONOPS. The JFC should develop clear mission statements and ensure they are understood by subordinates.

(3) **Commander's Planning Guidance.** JFCs guide the joint force's actions throughout planning and execution. However, the staff and component commanders typically expect the JFC to issue **initial guidance** soon after receipt of a mission or tasks from higher authority and provide more detailed **planning guidance** after the JFC approves an operational approach. This guidance is an important input to subsequent mission analysis, but the completion of mission analysis is another point at which the JFC may provide updated planning guidance that affects COA development.

(4) **Commander's Intent.** Commander's intent is the commander's clear and concise expression of what the force must do and the conditions the force must establish to accomplish the mission. It includes the purpose, end state, and associated risks. Commander's intent supports mission command and allows subordinates the greatest possible freedom of action. It provides focus to the staff and helps subordinate and supporting commanders act to achieve the commander's desired results without further orders once the operation begins, even when the operation does not unfold as planned. Successful commanders demand that subordinate leaders at all echelons exercise disciplined initiative and act aggressively and independently to accomplish the mission within the commander's intent. Subordinates emphasize timely decision making, understanding the higher commander's intent, and clearly identifying tasks to achieve desired objectives. Well-crafted commander's intent improves subordinates' situational awareness, which enables effective actions in fluid, chaotic situations.

(5) **CCIRs.** CCIRs are elements of friendly and enemy information the commander identifies as critical to timely decision making. They focus IM and help the JFC and staff assess the OE. The CCIR list is normally a product of mission analysis, and JFCs add, delete, and update CCIRs throughout an operation.

(6) **CONOPS.** The CONOPS, included in paragraph 3, (Execution) of the plan or order, describes how the JFC intends to integrate, synchronize, and phase actions of the joint force components and supporting organizations to accomplish the mission. CONOPS generally include potential branches and sequels. The CONOPS is typically a detailed extension of the operational approach, but incorporates modifications based on updated information and intelligence gained during planning as well as the JFC's approved COA. The staff writes (or graphically portrays) the CONOPS in sufficient detail so that subordinate and supporting commanders understand their mission, tasks, and other requirements and can develop their supporting plans accordingly. The CONOPS also provides the basis to develop the concept of fires, concept of intelligence operations, and theater logistics overview (TLO), which also are included in the final OPLAN or OPORD.

f. **Freedom of Action.** The JFC should maintain freedom of action throughout the operation. Freedom of action in the OA is linked to freedom to act beyond the OA. For example, operational reach—the distance and duration across which a joint force can successfully employ military capabilities—can extend far beyond the limits of a JFC's joint

Chapter II

operations area (JOA), and is inextricably tied to LOOs and the capacity and ability to throughput logistics to the point of destination. Consequently, the joint force must protect LOOs to ensure freedom of action. Attaining operational reach requires gaining and maintaining operational access in the face of enemy A2/AD capabilities and actions. Likewise, the C2 and intelligence functions depend on operations within the EMS and cyberspace. Losing the capability to operate effectively in the EMS and cyberspace can greatly diminish the JFC's freedom of action. While various actions (e.g., cybersecurity, cyberspace defense, joint electromagnetic spectrum operations [JEMSO], and the consideration of branches to current operations) contribute individually to freedom of action, operational design and joint planning are the processes that coherently link these actions. The JFC and staff should consider freedom of action from the outset of operational design and remain alert to indicators during operations that freedom of action is in jeopardy.

COMMON OPERATING PRECEPT

Ensure freedom of action.

Refer to JP 5-0, Joint Planning, *for more information on JPP. Refer to JP 3-09,* Joint Fire Support; *JP 3-09.3,* Close Air Support; *and JP 3-60,* Joint Targeting, *for more information on fires and joint fire support planning. Refer to JP 2-0,* Joint Intelligence, *and other intelligence series publications for more information on intelligence support and planning. Refer to JP 4-0,* Joint Logistics, *and other logistics series publications for more information on logistic planning. Refer to JP 3-12,* Cyberspace Operations, *for more information on cyberspace operations (CO). Refer to the* National Military Strategic Plan for Electronic Warfare, DOD Electromagnetic Spectrum (EMS) Strategy, *and JDN 3-16,* Joint Electromagnetic Spectrum Operations, *for more information on JEMSO/EMS superiority. Refer to JP 3-05,* Special Operations, *for more information on special operations planning.*

6. Assessment

a. Assessment is a continuous process that measures the overall effectiveness of employing joint force capabilities during military operations. It involves monitoring and evaluating the current situation and progress toward mission completion. Assessments can help determine whether a particular activity contributes to progress with respect to a set of standards or desired objective or end state. Assessments also help identify the current status of dynamic systems (e.g., weather, the economy, the political and security climate) and can help anticipate the future status.

b. DOD and its components use a wide range of assessment tools and methods. In peacetime and periods of conflict, assessments gauge the ability of the military instrument of national power to prepare for and respond to national security challenges described in Chapter I, "Fundamentals of Joint Operations." Leaders assess operations and activities across the levels of warfare and in all joint functions. At the strategic level, the CJCS conducts deliberate and continuous assessments such as the **Comprehensive Joint**

The Art of Joint Command

Assessment and the **CJCS's Readiness System,** respectively. Theater-strategic and operational-level assessments provide a methodology for joint commands and Services to adjust planning and execution to be more effective, match the dynamic OE, and better identify their risks and opportunities. At all levels, **staff estimates** are evaluations that assess factors in staff sections functional areas (e.g., intelligence, logistics). Staff estimates complement the overall operation assessment activity.

Refer to CJCSI 3100.01, Joint Strategic Planning System, for more information on CJCS assessments. Refer to JP 5-0, Joint Planning, for more information on staff estimates, integration of assessment during planning, and conducting operation assessment during execution.

    c. **Operation assessment** refers specifically to the process the JFC and staff use during planning and execution to measure progress toward accomplishing tasks, creating conditions or effects, and achieving objectives. Commanders continuously observe the OE and the progress of operations; compare the results to their initial visualization, understanding, and intent; and adjust planning and operations based on this analysis. Staffs monitor key factors that can influence operations and provide the commander information needed for decisions. Without mistaking level of activity for progress, commanders devise ways to update their understanding of the OE and assess their progress toward mission accomplishment. The fundamental aspects of assessment apply in all types of joint operations, although commanders and staffs may need to adjust the assessment process to fit the nature and requirements of a specific operation. In operations that do not include combat, assessments can be more complex.

    d. Assessment begins during mission analysis when the commander and staff consider what to measure and how to measure it. During further planning and preparation, the staff assesses the joint force's ability to execute the plan based on COAs that can meet planning objectives, available resources, and changing conditions in the OE. Throughout COA development, analysis, comparison, approval, and CONOPS finalization, the commander and staff devise the assessment process to incorporate in the plan and order. They will follow this process during plan development, subsequent refinement, adaptation, and execution. Key assessment indicators can be included in the **CCIR process** to provide timely support to the commander's planning and execution decisions.

    e. There is no uniform method by which joint forces assign management responsibilities for the assessment. The chief of staff's role will vary according to the commander's desires. Normally, the plans directorate of a joint staff (J-5), assisted by the training directorate of a joint staff (J-7) and intelligence directorate of a joint staff (J-2), develops the assessment plan during the planning process, while the operations directorate of a joint staff (J-3), assisted by the J-7 and J-2, coordinates assessment activities during execution. The assessment activity can also be overseen and managed by a separate analysis directorate or analysis element of an existing directorate if established. Various elements of the JFC's staff use assessment results to adjust both current operations and the J-5's future plans effort. Assessment is an entire staff effort requiring expertise and inputs across the staff. Formalizing assessment roles and responsibilities in each command is essential to an effective and efficient process.

Chapter II

f. During execution, assessment helps commanders decide whether and when to execute branches and sequels to align current and future operations with the mission and military end states. By including assessment key indicators within the CCIRs, the staff can better advise the commander whether the original operational approach is still valid.

g. Actions by a wide variety of entities affect military actions and objectives. These actors include interorganizational participants, the civilian population, neutral non-partner organizations in the JOA, and other countries outside the JOA in the GCC's AOR. Since assessment resources are limited, the commander must prioritize assessment activities. This typically requires collaboration with **interorganizational participants—preferably in a common, accepted** process—in the interest of unified action. Since most of these organizations are outside the JFC's authority, the JFC is responsible only for assessments of the activities of assigned, attached, and supporting military forces. Nevertheless, the JFC should grant some joint force organizations (e.g., civil affairs [CA] directorate or CMOC) authority to coordinate directly with organizations, such as DOS and Department of Homeland Security (DHS), and other CCMDs to facilitate effective integration and synchronization of assigned, attached, and supporting military forces, as well as timely and effective assessments by participants not under the JFC's authority.

h. **Assessment and Levels of Warfare**

(1) Interrelated and interdependent assessment occurs at all levels of warfare. Although each level of warfare may have a specific focus and a unique battle rhythm, together they form a hierarchical structure through which assessments interact (see Figure II-2). Typically, assessments at the theater-strategic and operational levels concentrate on broader tasks, effects, objectives, and military end state while assessments at the tactical level primarily focus on tasks, effects, and objectives. Properly focused analysis and collection at each level of warfare reduces redundancy and enhances the efficiency of the overall assessment.

(2) Operation assessment is most effective when supported and supporting plans and their assessments are linked. As Figure II-2 depicts, each level of assessment should be linked with adjacent levels, both to provide a conduit for guidance and provide information. For instance, assessment plans at the tactical level should delineate how they link to or support operational-level assessments. Similarly, guidance from the operational-level JFC should specify the relationship and mechanisms (e.g., tasks to subordinate organizations) by which tactical-level assessment data can be gathered and synthesized into the operational-level assessment.

(3) JFCs and their staffs consider assessment ways, means, and measures during planning, preparation, and execution. To optimize the assessment process given the scarcity of intelligence collection assets, JFCs and their staffs can include key assessment indicators in the CCIRs. This focuses assessment and collection at each level, reduces redundancy, and enhances the efficiency of the assessment process. At all levels, commanders and staffs develop operation assessment **indicators** to track progress toward mission accomplishment. An optimal method for developing indicators is to identify key assessment indicators associated with tasks, effects, objectives, and end states for inclusion

The Art of Joint Command

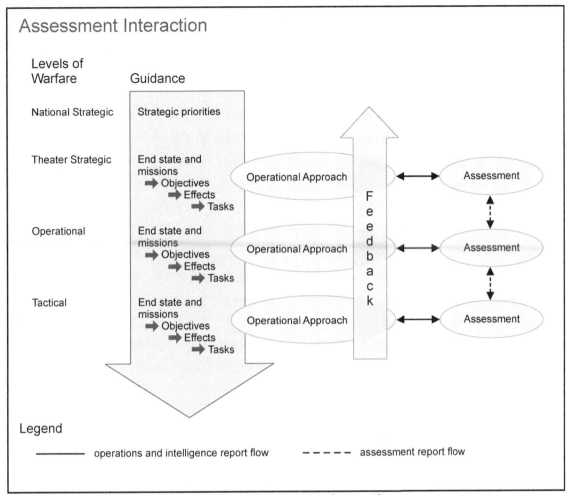

Figure II-2. Assessment Interaction

in the assessment design. The most critical indicators of progress or regression should also be included in CCIRs to guide the collection and assessment activity. These indicators include **measures of effectiveness** (MOEs) and **measures of performance** (MOPs). MOEs help answer the question, "Are we creating the effect(s) or conditions in the OE that we desire?" MOPs help answer the question, "Are we accomplishing tasks to standard?"

(4) Tactical-level assessment also uses MOEs and MOPs. Tactical tasks are often physical activities, but they can affect higher-level functions and systems. Tactical assessment may evaluate progress by phase lines; destruction of enemy forces; control of key terrain, peoples, or resources; and other tasks. Combat assessment evaluates the results of weapons engagement (of both lethal and nonlethal capabilities), and thus provides data for joint fires and the joint targeting processes at all levels. Combat assessment is composed of three related elements: battle damage assessment, munitions effectiveness assessment, and reattack recommendations or future targeting. Assessment of tactical results helps commanders determine progress at the operational and strategic levels and can affect operational and strategic targeting and engagement decisions. Tactical-level results provide JFCs comprehensive, integrated information to link tactical actions to operational and strategic objectives.

Chapter II

Refer to JP 5-0, Joint Planning, *for more information on operation assessment (e.g., integration of assessment design during the planning effort, roles and responsibilities, tenets of an effective assessment, the assessment process, and development and use of assessment indicators). Refer to JP 3-60,* Joint Targeting, *for more information on combat assessment.*

CHAPTER III
JOINT FUNCTIONS

> *"There are significant complexities to effectively integrating and synchronizing Service and combat support agency (CSA) capabilities in joint operations. These challenges are not new, and they present themselves with consistency. For example, simply getting the joint force to form and deploy in a coherent and desired manner requires integration of organization, planning, and communication capabilities and activities. But to fully employ the joint force in extensive and complex operations requires a much greater array of capabilities and procedures to help the commander and staff integrate and synchronize the joint force's actions. These types of activities and capabilities center on the commander's ability to employ the joint force and are grouped under one functional area called command and control. In a similar manner, many other functionally related capabilities and activities can be grouped. These groupings, we call joint functions, facilitate planning and employment of the joint force."*
>
> **Joint Publication 1, *Doctrine for the Armed Forces of the United States***

1. Introduction

a. This chapter discusses joint functions, related tasks, and key considerations. **Joint functions are related capabilities and activities grouped together to help JFCs integrate, synchronize, and direct joint operations.** Functions that are common to joint operations at all levels of warfare fall into **six basic groups—C2, intelligence, fires, movement and maneuver, protection, and sustainment.** Some functions, such as C2 and intelligence, apply to all operations. Others, such as fires, apply as the JFC's mission requires. A number of subordinate tasks, missions, and related capabilities help define each function, and some could apply to more than one joint function.

b. The joint functions reinforce and complement one another, and **integration across the functions is essential to mission accomplishment.** For example, joint fires can enhance the protection of a joint security area (JSA) by dispersing or disrupting enemy assets threatening the JSA. In any joint operation, the JFC can choose from a wide variety of joint and Service capabilities and combine them in various ways to perform joint functions and accomplish the mission. Plans describe how the JFC uses military capabilities (i.e., organizations, people, and systems) to perform tasks associated with each joint function. However, forces and other assets are not characterized by the functions for which the JFC is employing them. Individual Service capabilities can often support multiple functions simultaneously or sequentially while the joint force is executing a single task. For example, aviation assets routinely support all six functions in a single combat operation. Just as component commanders integrate activities across functions to accomplish tasks and missions, the JFC and staff do likewise for the joint force. Various factors complicate the JFC's integration challenge, such as competing demands for high-priority capabilities and the fact that joint force components have different function-oriented approaches, procedures, and perspectives. JFCs and USG interagency partners synchronize, coordinate, and integrate military operations with the activities of interorganizational participants to achieve unity of effort toward US objectives. Military

Chapter III

forces support the USG lead agency, department, or organization, which is usually DOS for overseas operations, and DHS for domestic operations.

c. JFCs and staffs integrate, synchronize, employ, and assess a wide variety of information-related capabilities (IRCs) within and across joint functions, in concert with other actions to influence a target audience's decision making while protecting our own. IRCs constitute tools, techniques, or activities employed through the information environment that can be used to create effects, accomplish tasks, or achieve specific objectives at a specific time and place. IRCs reinforce and complement one another, and their integration is essential to mission accomplishment. Integration and synchronization of and across IRCs enables many aspects of joint operations. The JFC should ensure the staff coordinates between IO, CA, public affairs (PA), and defense support to public diplomacy (DSPD) to enable effective CCS efforts. The JFC's objectives require early detailed IO staff planning, coordination, and deconfliction between the USG and PN efforts within the OA and affected areas, in order to effectively synchronize and integrate IRCs to create coordinated effects. Information is a foundational element of joint operations and these related capabilities and activities help JFCs integrate, synchronize, and direct joint operations. Coupled with the requirements of the current operational environment, JFCs use information more as a joint function. For this reasons, the joint force is considering adding information as a seventh joint function.

See JP 3-13, Information Operations, for more information on IO.

2. Command and Control

a. C2 encompasses the exercise of authority and direction by a commander over assigned and attached forces to accomplish the mission. The JFC provides operational vision, guidance, and direction to the joint force. The **C2 function** encompasses a number of tasks, including:

(1) Establish, organize, and operate a joint force HQ.

(2) Command subordinate forces.

(3) Prepare, modify, and publish plans, orders, and guidance.

(4) Establish command authorities among subordinate commanders.

(5) Assign tasks, prescribe task performance standards, and designate OAs.

(6) Prioritize and allocate resources.

(7) Manage risk.

(8) Communicate and maintain the status of information across the staff, joint force, and with the public as appropriate.

(9) Assess progress toward accomplishing tasks, creating conditions, and achieving objectives.

(10) Coordinate and control the employment of joint lethal and nonlethal capabilities.

(11) Coordinate, synchronize, and when appropriate, integrate joint operations with the operations and activities of other participants.

(12) Ensure the flow of information and reports to higher authority.

b. **Command** includes both the authority and responsibility to use resources to accomplish assigned missions. Command at all levels is the art of motivating and directing people and organizations to accomplish missions. The C2 function supports an efficient decision-making process. Timely intelligence **enables commanders to make decisions and execute those decisions more rapidly and effectively than the enemy.** This decreases risk and allows the commander more control over the timing and tempo of operations.

c. **Command Authority.** JFCs exercise various command authorities (i.e., combatant command [command authority] {COCOM}, OPCON, tactical control [TACON], and support) delegated to them by law or senior leaders and commanders over assigned and attached forces. **Command relationship** is a term that describes the relationships established through the designation of these authorities. JP 1, *Doctrine for the Armed Forces of the United States,* provides details on each authority, and Figure III-1 summarizes their relationships. Unity of command among US forces is maintained through the application of the various command relationships shown below.

(1) COCOM is the nontransferable command authority established by Title 10, United States Code (USC), Section 164, exercised only by commanders of unified or specified CCMDs unless otherwise directed by the President or SecDef. COCOM, which cannot be delegated, is the authority of a CCDR to perform those functions of command over assigned forces involving organizing and employing commands and forces; assigning tasks; designating objectives; and giving authoritative direction over all aspects of military operations, joint training, and logistics necessary to accomplish the missions assigned to the command. COCOM should be exercised through the commanders of subordinate organizations. Normally, this authority is exercised through subordinate JFCs and Service and/or functional component commanders. COCOM provides full authority to organize and employ commands and forces as the CCDR considers necessary to accomplish assigned missions. During crisis response and combat, or where critical situations require changing the normal logistic process, the CCDRs' directive authority for logistics (DAFL) enables them to use all logistic capabilities of all forces assigned and attached to their commands as necessary to accomplish their mission. Under peacetime conditions, the CCDR will exercise logistic authority consistent with the peacetime limitations imposed by legislation, DOD policy or regulations, budgetary considerations, local conditions, and other specific conditions prescribed by SecDef or CJCS.

Chapter III

Command Relationships Synopsis

Combatant Command (Command Authority)

(Unique to Combatant Commander)

- Planning, programming, budgeting, and execution process input
- Assignment of subordinate commanders
- Relationships with Department of Defense agencies
- Directive authority for logistics

Operational control when delegated

- Authoritative direction for all military operations and joint training
- Organize and employ commands and forces
- Assign command functions to subordinates
- Establish plans and requirements for intelligence, surveillance, and reconnaissance activities
- Suspend subordinate commanders from duty

Tactical control when delegated

Local direction and control of movements or maneuvers to accomplish mission

Support relationship when assigned

Aid, assist, protect, or sustain another organization

Figure III-1. Command Relationships Synopsis

(2) **OPCON** is inherent in COCOM and may be delegated within the command. OPCON is command authority that may be exercised by commanders at any echelon at or below the level of CCMD to perform those functions of command over subordinate forces. It involves organizing and employing commands and forces, assigning tasks, designating objectives, and giving authoritative direction necessary to accomplish the mission. OPCON includes authoritative direction over all aspects of military operations and joint training necessary to accomplish missions assigned to the command. This authority should be exercised through the commanders of subordinate organizations, normally through subordinate JFCs and Service and/or functional component commanders. OPCON normally provides full authority to organize commands and forces and to employ those forces as the commander exercising OPCON considers necessary to accomplish assigned missions; it does not, in and of itself, include DAFL or matters of administration, discipline, internal organization, or unit training.

(3) TACON is inherent in OPCON. TACON is an authority over assigned or attached forces or commands, or military capability or forces, made available for tasking. It is limited to the detailed direction and control of movements or maneuvers within the

> **COMMAND RELATIONSHIPS DURING OPERATION IRAQI FREEDOM**
>
> In December 2002, representatives from United States Central Command (USCENTCOM) and United States European Command (USEUCOM) met in Stuttgart, Germany to discuss Operation IRAQI FREEDOM (OIF). The two broad issues were organizing the operational area and coordinating the command relationships for all OIF phases.
>
> The USCENTCOM OIF theater of operations would, by necessity, cross the Unified Command Plan (UCP)-designated USCENTCOM and USEUCOM areas of responsibility (AORs) boundary. Specifically, the land and airspace of Turkey was recognized for its potential to contribute to opening a northern line of operation.
>
> Discussions over the potential options for organizing the OIF operational area led to an agreement not to request a temporary change in the UCP modifying the AORs, but to rely on the establishment of appropriate command relationships between the two combatant commanders (CCDRs).
>
> Discussions over the potential command and control options led to the decision to establish a support relationship between USCENTCOM (supported) and USEUCOM (supporting). This relationship was established by the Secretary of Defense. It enabled the development of coherent and supporting campaign plans.
>
> In the campaign plan, USEUCOM retained tactical control for the coordination and execution of operational movement (i.e., reception, staging, onward movement, and integration); intelligence, surveillance, and reconnaissance; logistic and personnel support; and protection in support of USCENTCOM forces transiting the USEUCOM AOR, specifically Turkey. Once USCENTCOM-allocated joint forces were positioned and prepared to cross the Turkish–Iraqi border (to commence offensive operations) operational control would be given to USCENTCOM.
>
> Maintaining UCP AOR boundaries and the establishment of an umbrella support relationship between the CCDRs with conditional command authorities exercised over the participating forces based on their readiness and operation phase provided a workable solution to the integration and employment of joint forces on the boundary of two AORs.
>
> **Various Sources**

OA necessary to accomplish assigned missions or tasks assigned by the commander exercising OPCON or TACON of the attached force. TACON may be delegated to, and exercised at, any level at or below the level of CCMD. TACON provides sufficient authority for controlling and directing the application of force or tactical use of combat support assets within the assigned mission or task. Commanders may delegate TACON to subordinate commanders at any echelon at or below CCMD and may exercise TACON over assigned or attached forces or military capabilities or forces made available for tasking. TACON does not provide organizational authority or authoritative direction for

Chapter III

administrative and logistic support or discipline (Uniform Code of Military Justice authority); the commander of the parent unit continues to exercise those responsibilities unless the establishing directive specifies otherwise. Except for special operations forces (SOF), functional component commanders typically exercise TACON over military capabilities or forces made available for tasking.

(4) **Support.** Establishing support relationships between components (as described in JP 1, *Doctrine for the Armed Forces of the United States*) is a useful option to accomplish needed tasks. **The JFC can establish support relationships among all functional and Service component commanders,** such as for the coordination of operations in depth involving the joint force land component commander (JFLCC) and the joint force air component commander (JFACC). Within a joint force, the JFC may designate more than one supported commander simultaneously, and components may simultaneously receive and provide support for different missions, functions, or operations. For instance, a joint force special operations component commander (JFSOCC) may be supported for a direct-action mission while simultaneously supporting a JFLCC for a raid. Similarly, a joint force maritime component commander (JFMCC) may be supported for a sea control mission while simultaneously supporting a JFACC to achieve control of the air throughout the OA.

(5) **Other authorities** granted to commanders, and to subordinates as required, include **administrative control, coordinating authority, directive authority for CO, and direct liaison authorized.** JP 1, *Doctrine for the Armed Forces of the United States,* outlines the specific details for each command relationship.

> **COMMON OPERATING PRECEPT**
>
> **Avoid combining capabilities where doing so adds complexity without compensating advantage.**

(6) The perceived benefits of operations by joint forces do not occur naturally just by virtue of command relationships. The integration necessary for effective joint operations requires explicit effort; can increase operational complexity; and will require additional training, technical and technological interoperability, liaison, and planning. Although effectiveness is typically more important than efficiency in joint operations, the JFC and component commanders must determine when the potential benefits of joint integration cannot compensate for the additional complicating factors. Synergy is a means to greater operational effectiveness, not an end in itself. The joint operations principle of simplicity is always a key consideration.

d. **Control is inherent in command.** To control is to manage and direct forces and functions consistent with a commander's command authority. Control of forces and functions helps commanders and staffs compute requirements, allocate means, and integrate efforts. Control is necessary to determine the status of organizational effectiveness, identify variance from set standards, and correct deviations from these standards. Control permits commanders to acquire and apply means to support the mission and develop specific instructions from general guidance. Control provides the means for commanders to maintain freedom of action, delegate authority, direct operations from any location, and integrate and synchronize actions throughout the OA. Ultimately, it provides commanders a means to measure, report, and correct performance.

e. **Area of Operations (AO) and Functional Considerations**

(1) **C2 in an AO. The land and maritime force commanders are the supported commanders within their designated AOs.** Through C2, JFLCCs and JFMCCs integrate and synchronize movement and maneuver with intelligence, fires, protection, and sustainment and the supporting IRCs. To facilitate this integration and synchronization, they have the authority to designate target priority, effects, and timing of fires within their AOs.

(a) Synchronization of efforts within land or maritime AOs with theater- and/or JOA-wide operations is of particular importance. To facilitate synchronization, the JFC establishes priorities that will guide or inform execution decisions throughout the theater and/or JOA, including within the land, maritime, and SOF commander's AOs. **The JFACC is normally the supported commander for the JFC's overall air interdiction effort, while JFLCCs and JFMCCs are supported commanders for interdiction in their AOs.**

(b) In coordination with JFLCCs and JFMCCs, other commanders tasked by the JFC to execute theater- or JOA-wide operations have the latitude to plan and execute them within land and maritime AOs. Commanders executing such operations within a land or maritime AO must coordinate the operation with the appropriate commander to avoid adverse effects and friendly fire incidents. **If planned operations would have adverse impact within a land or maritime AO, the commander assigned to execute the JOA-wide functions must readjust the plan, resolve the issue with the land or maritime component commander, or consult with the JFC for resolution.**

For additional guidance on C2 of land or maritime operations, refer to JP 3-31, Command and Control for Joint Land Operations, *and JP 3-32,* Command and Control for Joint Maritime Operations.

(2) **C2 of Space Operations.** A supported JFC normally designates a space coordinating authority (SCA) to coordinate joint space operations and integrate space capabilities. Based on the complexity and scope of operations, the JFC can either retain SCA or designate a component commander as the SCA. The JFC considers the mission, nature, and duration of the operation; preponderance of space force capabilities made available; and resident C2 capabilities (including reachback) when selecting the appropriate option. The **SCA** determines the joint force's space requirements, coordinates and integrates space capabilities in the OA, and plans and assesses joint space operations. The SCA is normally supported by either assigned or attached embedded space personnel. There are established doctrinal processes for articulating requirements for space force enhancement products. These processes are specifically tailored to the functional area they support and result in prioritized requirements. Thus, the SCA typically has no role in

Chapter III

prioritizing the JFC's day-to-day space force enhancement requirements. The SCA gathers operational requirements that may be satisfied by space capabilities and facilitates the use of established processes by joint force staff to plan and conduct space operations. Following coordination, the SCA provides the JFC a prioritized list of recommended space requirements based on joint force objectives. To ensure prompt and timely support, the supported CCDR and Commander, United States Strategic Command (CDRUSSTRATCOM), may authorize direct liaison between the SCA and applicable United States Strategic Command (USSTRATCOM) component(s). Joint force component commands should communicate their requirements to the SCA or designated representative to ensure all space activities are properly integrated and synchronized.

Refer to JP 3-14, Space Operations, *for detailed guidance on C2 of space operations.*

(3) **C2 of Joint Air Operations.** The JFC normally designates a JFACC to establish unity of command and unity of effort for joint air operations. The JFC delegates the JFACC the authority necessary to accomplish assigned missions and tasks. The JFC may also establish support relationships between the JFACC and other components to facilitate operations. The JFACC conducts joint air operations IAW the JFC's intent and CONOPS. The JFC may designate the JFACC as the supported commander for strategic attack; air interdiction; personnel recovery (PR); and airborne intelligence, surveillance, and reconnaissance (ISR) (among other missions). As such, the JFACC plans, coordinates, executes, and assesses these missions for the JFC. Additionally, the JFC also normally designates the JFACC as the area air defense commander (AADC) and airspace control authority (ACA) because the three functions are integral to one another and require an OA-wide perspective. When appropriate, the JFC may designate a separate AADC or ACA. In those joint operations where separate commanders are designated, close coordination is essential for unity of effort, prevention of friendly fire incidents, and deconfliction of joint air operations.

(a) **ACA.** The JFC is ultimately responsible for airspace control within the OA, but normally delegates the authority to the ACA. The ACA, in conjunction with the Service and functional components, coordinates and integrates the use of the airspace and develops guidance, techniques, and procedures for airspace control and for units operating within the OA. The ACA establishes an airspace control system (ACS) that is responsive to the JFC's needs, integrates the ACS with the HN, and coordinates and deconflicts user requirements. The airspace control plan (ACP) and airspace control order (ACO) express how the airspace will be used to support mission accomplishment. The ACA develops the ACP, coordinates it with other joint force component commanders, and, after JFC approval, distributes it throughout the OA and to all supporting airspace users. The ACP establishes guidance for the development of the ACS and distribution of the ACO. The ACA publishes the ACO to maximize the combat effectiveness of the joint force and to support mission accomplishment IAW JFC priorities. See JP 3-52, *Joint Airspace Control,* and JP 3-30, *Command and Control of Joint Air Operations,* for more information.

(b) **AADC.** The AADC is responsible for defensive counterair (DCA) (which includes both air and missile threats) operations. The AADC must identify those volumes of airspace and control measures that support and enhance DCA operations, identify required airspace management systems, establish procedures for systems to operate within the airspace, and ensure they are incorporated into the ACS. During complex operations or campaigns conducted in a large theater of operations, the AADC may recommend, and the JFC may choose, to divide the JOA into separate air defense regions, each with a regional air defense commander who could be delegated responsibilities and decision-making authority for DCA operations within the region. See JP 3-01, *Countering Air and Missile Threats,* for more information.

(4) **C2 of Joint Maritime Operations.** JFCs establish maritime AOs to achieve unity of command over the execution of maritime component operations involving the interrelated employment of joint air, surface, and subsurface forces. The maritime AO should be of sufficient size to allow for movement, maneuver, and employment of weapons systems; effective use of warfighting capabilities; and provide operational depth for sustainment and force protection. The JFMCC is the supported commander for operations within the JFC-designated maritime AO. The AADC will normally establish a congruent air defense region, covering the open ocean and littorals, with a regional air defense commander, who is delegated decision-making authority for counterair operations within the region. The maritime regional air defense commander and AADC will coordinate to ensure that the JFACC can accomplish theater-wide responsibilities assigned by the JFC.

(5) **C2 of Joint CO**

(a) Commanders conduct CO to retain freedom of maneuver in cyberspace, accomplish the JFC's objectives, deny freedom of action to enemies, and enable other operational activities. Some of the capabilities the JFC may employ to enable CO include significant portions of JEMSO, C2, intelligence collection, navigation warfare, and some space mission areas.

(b) The CCDR will organize a staff capable of planning, synchronizing, and controlling CO in support of their assigned mission. Each CCMD supports subordinate JFCs through their CO supporting staff and assigned United States Cyber Command (USCYBERCOM) cyberspace support element (CSE). Clearly established command relationships are crucial for ensuring timely and effective employment of cyberspace capabilities. While USCYBERCOM exercises directive authority for CO, it coordinates all actions with the affected CCMDs through their CSEs to facilitate unity of effort and mission accomplishment. The CCMD coordinates and deconflicts all cyberspace missions in the AOR with other operations, including nationally tasked actions and the cyberspace actions initiated in the CCMD. The CCMD coordinates and integrates cyberspace capabilities in the AOR and has primary responsibility for joint CO planning, to include determining cyberspace requirements within the joint force.

(c) Leveraging USCYBERCOM capacity, through the CSE, the CCMD will integrate cyberspace capabilities into plans, deconflict and synchronize supporting cyberspace fires, prepare the OE, and conduct operational assessments and readiness functions. Additionally, in partnership with USCYBERCOM, the CCMD will coordinate regionally with interagency and allied participants as necessary. They will integrate cyberspace command, planning, operations, intelligence, targeting, and readiness

Chapter III

processes for creating cyberspace effects with the CCDR's plans and operations through the three CO missions: offensive CO, defensive cyberspace operations (DCO), and DODIN operations. The CCMD may be supported by assigned or attached embedded cyberspace forces. CCMDs will collaborate with their counterparts in the other CCMDs and with USCYBERCOM when initiating cyberspace actions with possible effects outside their AOR.

For guidance on C2 of cyberspace forces, refer to JP 3-12, Cyberspace Operations.

(d) **DODIN Operations**

1. The DODIN is the set of information capabilities and associated processes to collect, process, store, disseminate, and manage information on demand for warfighters, policy makers, and support personnel, whether interconnected or stand-alone. It includes owned and leased communications and computing systems and services, software (including applications), data, security services, other associated services, and national security systems. DODIN operations design, build, configure, secure, operate, maintain, and sustain DOD information networks. Since all actions taken for the sake of performance impact security and vice versa, DODIN operations and DCO internal defensive measures are intrinsically linked activities. DODIN operations are the means by which DOD manages the flow of information over its information networks. The purpose of DODIN operations is assured system and network availability, assured information protection, and assured information delivery, which protect and maintain freedom of action for DOD missions within cyberspace. DODIN operations require centralized coordination because they have the potential to impact the integrity and operational readiness of the DODIN; however, mission execution is generally decentralized. The DODIN supports all military operations by enabling DOD, multinational, international, nongovernmental, and interagency mission partners to securely and seamlessly share required information. The aggregate effect of DODIN operations activities establishes the security framework on which all DOD missions ultimately depend.

2. CO are enabled by the DODIN, and DODIN operations are CO missions. CDRUSSTRATCOM is the supported commander for global DODIN operations and synchronizes planning for other CO. USCYBERCOM directs security, operations, and defense of the DODIN. It is responsible for conducting full-spectrum military CO to enable US freedom of action in cyberspace and enable actions throughout the OE and deny the same to our adversaries and enemies. CDRUSSTRATCOM, through Commander, USCYBERCOM, also is a supporting commander for regional DODIN operations and provides support to affected CCMDs, Services, and DOD agencies. Other USG departments and agencies may also provide support per intra-governmental agreements.

For additional guidance on DODIN operations, refer to JP 6-0, Joint Communications System, *and JP 3-12,* Cyberspace Operations.

(6) **C2 of JEMSO.** JEMSO are those activities consisting of electronic warfare (EW) and joint EMS management operations used to exploit, attack, protect, and manage

Joint Functions

the electromagnetic environment (EME) to achieve the commander's objectives. The electronic warfare control authority (EWCA), the senior electronic attack authority in the OA, develops guidance to execute electronic attack (EA) on behalf of the JFC. EWCA can either be retained by the JFC or executed by the JFC's designated representative, normally the J-3. When so authorized, the J-3 will have primary staff responsibility to plan, coordinate, integrate, and monitor execution of joint force EW operations. Frequency allotment and assignment authority is normally delegated to the joint frequency management office, who may further delegate this authority to facilitate decentralization and to provide components with the maximum latitude and flexibility in support of combat operations. The J-3/EWCA must be cognizant of the effect EW and JEMSO may have on CO since some networks have free-space radio frequency links and therefore can be affected by EW operations and JEMSO needs to take CO into consideration.

For additional guidance on the communications and intelligence systems support and JEMSO, refer to JP 2-01, Joint and National Intelligence Support to Military Operations; *JP 6-0,* Joint Communications System; *JP 6-01,* Joint Electromagnetic Spectrum Management Operations; *JDN 3-13,* Joint Electromagnetic Spectrum Operations, *and JP 3-13.1,* Electronic Warfare.

    f. **C2 System.** JFCs exercise authority and direction through a C2 system, which consists of the facilities; equipment; communications; staff functions and procedures; and personnel essential for planning, preparing for, monitoring, and assessing operations. The C2 system must enable the JFC to maintain communication with higher, supporting, and subordinate commands in order to control all aspects of current operations while planning for future operations.

    (1) The joint force staff is the linchpin of the C2 system, since the JFC understands, plans, directs, and controls most aspects of operations through the staff's expertise and efforts.

    (2) **Liaison** is an important aspect of C2. Commanders may exchange liaison teams or individuals between higher, supporting, and subordinate commands as required. Liaison personnel generally represent the interests of the sending commander to the receiving commander, but can greatly promote understanding of the commander's intent at both the sending and receiving HQ; they should be assigned early during joint planning. Liaison officers (LNOs) from supporting to supported commanders are particularly essential in determining needs and coordinating supporting actions.

    (3) **Control and Coordination Measures.** JFCs establish various maneuver and movement control, airspace coordinating, and fire support coordination measures to facilitate effective joint operations. These measures include boundaries, phase lines, objectives, coordinating altitudes to deconflict air operations, air defense areas, OAs, submarine operating patrol areas, no-fire areas, and others as required.

For additional guidance on C2 of air operations, refer to JP 3-30, Command and Control of Joint Air Operations. *For additional guidance on control and coordination measures, refer to JP 3-09,* Joint Fire Support, *and JP 3-52,* Joint Airspace Control. *See Military*

Chapter III

Standard-2525, Department of Defense Interface Standard Joint Military Symbology, *for additional guidance on the use and discussion of graphic control measures and symbols for the joint force.*

(4) **Communications and ISR systems** provide communications, intelligence, targeting data, and missile warning. The precision, speed, and interoperability with which these systems operate improve access to the information available to all command levels, thereby enhancing a common perspective of the OE. Effective command at varying operational tempos requires timely, reliable, secure, interoperable, and sustainable communications. Communications and ISR planning increases options available to JFCs by providing the communications sensor systems necessary to collect, process, store, protect, and disseminate information at decisive times. These communications and sensor systems permit JFCs to exploit tactical success and facilitate future operations.

(a) **Communications System Planning.** The purpose of the joint communications system is to assist the JFC in C2 of military operations. Effective communication system planning is essential for effective C2 and integration and employment of the joint force's capabilities. The mission and structure of the joint force determine specific information flow and processing requirements. These requirements dictate the general architecture and specific configuration of the communications system. Therefore, communications system planning must be integrated and synchronized with joint planning. Through effective communications system planning, the JFC is able to apply capabilities at the critical time and place for mission success. Communications system planning considers, and when appropriate, accommodates communications links with relevant commanders and leaders and their C2 (or equivalent) centers. Interoperability, foreign disclosure authorities, information sharing, and communications security planning with these stakeholders is essential to ensure secure communications and protect sensitive information. Routine communications and backup systems may be disrupted, and civil authorities might have to rely on available military communications equipment. Communications system planning also must consider termination of US involvement and procedures to transfer communications system control to another agency such as the United Nations (UN). Planning should consider that it may be necessary to leave some communications resources behind to continue support of the ongoing effort.

(b) Joint communications system planning and management involves the employment and technical control of assigned communications systems. Communications system planning allows the planners to maintain an accurate and detailed status of the network down to the modular level. Essential elements of the communications system are driven by the mission and determined by the C2 organization and location of forces available to the JFC. Specific command relationships and the organization of units and staffs drive the interconnecting communications methods and means. The communications system must support and provide for assured flow of information to and from commanders at all levels.

(c) Communications system planners ensure the organization's communications network can facilitate a rapid, unconstrained flow of information from its source through intermediate collection and processing nodes to its delivery to the user.

Typically, the combined system will provide voice, data, and video communications. Building the communications system to support the JFC requires knowledge of the joint force organization, the commander's CONOPS, communications available, and how they are employed. The ability to command, control, and communicate with globally deployed forces is a key enabler for protection of US national interests and, as such, is also a key target for adversaries. Thus, it is essential to consider risk and mitigation measures when developing the plan. Key planning considerations include protecting the DODIN, which requires cybersecurity and cyberspace defense measures to protect, detect, respond to, restore, and react to shield and preserve information and information systems. A related consideration is to ensure the aggregation of data within the communications systems does not compile information that must be protected at a higher level of security than the system provides (e.g., classified information on an unclassified system).

(d) The communications system being planned is the primary means through which intelligence flows to the JFC and throughout the OE. Communications system planning must be conducted in close coordination with the intelligence community to identify specialized equipment and dissemination requirements for some types of information.

(e) Homeland Security and Defense Communications System Planning. DOD contributes to homeland security through its military missions overseas, homeland defense (HD), and defense support of civil authorities (DSCA). The disparity of communications systems, use of allocated bandwidth (both civilian and military), and limited interoperable systems hinders the capability of collaborative incident management and response in the US. Commanders and communications system planners need to consider the detailed planning and analysis to determine US-based communications system requirements in support of federal, state, and local agencies.

(f) DOD intelligence component capabilities, resources, and personnel may not be used for activities other than foreign intelligence or counterintelligence (CI), unless SecDef specifically approves that use. Also, requests for direct DOD support to civilian law enforcement agencies (LEAs) are closely reviewed and processed separately for approval to ensure compliance with the Posse Comitatus Act. When approved, use of intelligence capabilities for domestic non-intelligence activities is limited to **incident awareness and assessment.** All incident awareness and assessment support within the US is subject to USG intelligence oversight regulations and Department of Defense Directive (DODD) 5240.01, *DOD Intelligence Activities*.

g. **CCIRs**

(1) **CCIRs are elements of information that the commander identifies as critical to timely decision making.** CCIRs focus IM and help the commander assess the OE and identify decision points during operations. **CCIRs belong exclusively to the commander.** The CCIR list is normally short so that the staff can focus its efforts and allocate scarce resources. But the CCIR list is not static; JFCs add, delete, adjust, and update CCIRs throughout planning and execution based on the information they need to

Chapter III

make decisions. At a minimum, CCIRs should be reviewed and updated throughout plan development, refinement, and adaptation, and during each phase of order execution.

(2) **Categories.** Priority intelligence requirements (PIRs) and friendly force information requirements (FFIRs) constitute the total list of CCIRs.

(a) **PIRs** are designated by the commander to focus information collection on the enemy or adversary and the OE to provide information required for decision making. All staff sections can recommend potential PIRs that may support the JFC's decision-making process. However, the J-2 consolidates the staff's recommended PIRs to the commander. PIRs are continuously updated in synchronization with the commander's decision points. PIRs are periodically reviewed to support plan refinement and adaptation based on the OE and prior to each execution phase transition to ensure the PIRs remain relevant to the commander's anticipated decision points.

Refer to JP 2-0, Joint Intelligence, *for more information on PIRs.*

(b) **FFIRs** focus on information the JFC must have to assess the status of the friendly force and supporting capabilities. All staff sections can recommend potential FFIRs that meet the JFC's guidance. The J-5 typically consolidates FFIR nominations and provides staff recommendation to the commander during planning prior to execution. During execution, the joint force J-3 consolidates these nominations and provides the recommendation for FFIRs that relate to current operations. The J-5 consolidates nominations and recommends FFIRs related to the future plans effort. JFC-approved FFIRs are automatically CCIRs.

h. **Battle Rhythm.** The HQ battle rhythm is its daily operations cycle of briefings, meetings, and report requirements. A stable battle rhythm facilitates effective decision making, efficient staff actions, and management of information within the HQ and with higher, supporting, and subordinate HQ. The commander and staff should develop a battle rhythm that minimizes meeting requirements while providing venues for command and staff interaction internal to the joint force HQ and with subordinate commands. Joint and component HQ's battle rhythms should be synchronized to accommodate operations in multiple time zones and the battle rhythm of higher, subordinate, and adjacent commands. Other factors such as planning, decision making, and operating cycles (i.e., intelligence collection, targeting, and joint air tasking cycle) influence the battle rhythm. Further, meetings of the staff organizations must be synchronized. The chief of staff normally manages the joint force HQ's battle rhythm. When coordinating with other USG departments and agencies, the joint force HQ should consider that those organizations often have limited capabilities and restricted access to some information.

i. **Building Shared Understanding.** Decisions are the most important products of the C2 function, because they guide the force toward objectives and mission accomplishment. Commanders and staff require not only information to make these decisions, but also the knowledge and understanding that results in the wisdom essential to sound decision making (Figure III-2).

Joint Functions

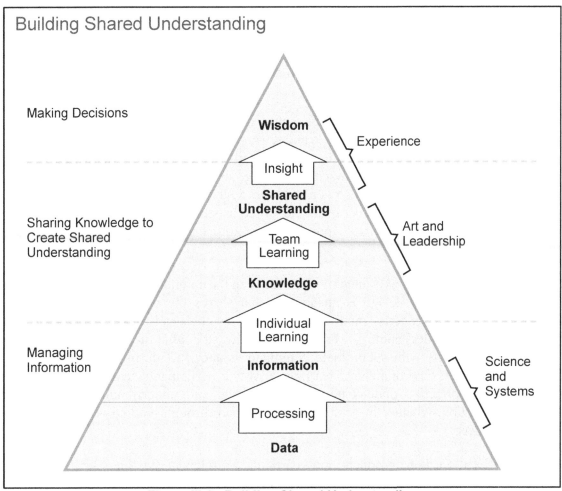

Figure III-2. Building Shared Understanding

(1) **IM.** IM is an essential process that receives, organizes, stores, controls, and secures an organization's wide range of data and information. It facilitates availability to relevant users to develop understanding through knowledge sharing, while concurrently preventing inadvertent disclosure of sensitive or proprietary information. IM is more than an information technology process; IM involves numerous decision support tools intimately integrated with the CCIRs. IM includes hard-copy decision support matrixes for the chief of staff and staff officers, wall-charts with PIR and FFIR statuses, report logs, call logs, video feeds, and information storage directives. IM is important for the commander's battle rhythm and the development and sharing of information to increase both individual and collective knowledge. Effective IM improves the speed and accuracy of information flow and supports execution through reliable communications. The process is used to manage the organization's information resources and optimize access to information by all who need it. As the key joint force staff integrator, the chief of staff may be responsible for managing the IM process, while the communications system directorate of a joint staff ensures the operation and connectivity of the supporting C2 communications systems and processes. Many joint HQ will also have an IM officer and an IM plan. HQ may also form a joint IM board to serve as a focal point for information oversight and coordination. HQ's standard operating procedures (SOPs) will normally

Chapter III

describe specific IM procedures and responsibilities. IM feeds the development and sharing of knowledge-based information products.

For further guidance on IM, refer to JP 6-0, Joint Communications System, *and JP 3-33,* Joint Task Force Headquarters.

(2) **Knowledge sharing** complements the value of IM with processes to create an organizational culture that encourages and rewards **team learning** activities **to facilitate shared understanding.** Knowledge sharing involves taking a deliberate, cross-organizational and functional approach to gaining, sharing, and maintaining knowledge that facilitates understanding, which is necessary to have an advantage in the OE. While information can be collected, processed, and stored as structured or unstructured content, such as in reports and databases, individuals acquire knowledge through a cognitive process.

(a) The free exchange of ideas between the commander and staff that should typify early operational design is an activity that shares the individual knowledge of numerous functional experts, modifies and increases their collective knowledge, and promotes shared understanding. In a similar way, the after action sessions that a commander conducts with subordinate commanders and staff during and following an operation create an environment of learning in which participants share knowledge and increase their collective understanding. Knowledge and understanding occur better through interaction, whether in person or virtual, than through reading and assimilating various products.

(b) Certain products are particularly relevant to knowledge sharing. **Commander's intent** is a knowledge-based product that commanders use to share their insight and direction with the joint force. The intent creates shared purpose and understanding, provides focus to the staff, and helps subordinate and supporting commanders act to achieve objectives without further orders, even when operations do not unfold as planned. Likewise, lessons-learned databases are knowledge-based products that help users avoid previous mistakes and adopt proven best practices. These databases exemplify how the marriage of IM and decision-support processes can improve future operations by sharing knowledge gained through experience.

(3) Specific IM, knowledge-sharing, and other collaborative processes and products vary across joint commands based on the commander's needs and preferences. The HQ SOP should specify these procedures and products and describe the relationship between IM, knowledge sharing, and the HQ battle rhythm.

(4) **Collaboration**

(a) Effective collaboration enhances C2 by sharing knowledge and building shared understanding. Although the value of face-to-face interaction is indisputably preferred, capabilities that improve long-distance, asynchronous collaboration among dispersed forces can enhance both planning and execution of joint operations. One has to consider, however, the added risk as long-distance collaboration may create a critical

Joint Functions

vulnerability that an enemy or adversary can exploit. These capabilities not only can improve efficiency and common understanding during routine, peacetime interaction among participants, they can also enhance combat effectiveness during time-compressed operations associated with both combat and noncombat operations.

(b) A collaborative environment is one in which participants are encouraged to use critical thinking; solve problems; and share information, knowledge, perceptions, ideas, and concepts in a spirit of mutual cooperation that extends beyond the requirement to coordinate with others. In joint operations, commanders and staffs tend to collaborate due to an established common purpose. However, collaboration can be enhanced when personnel leverage social networks, establish trusted relationships, and share knowledge. This is particularly important in relationships with interorganizational participants, since their objectives, perceptions, and desired results will not always coincide with the military's.

(c) **Collaboration Capabilities.** Collaboration capabilities can enable planners and operators worldwide to build a plan without being collocated. Collaboration also provides planners with a "view of the whole" while working on various sections of a plan, which helps them identify and resolve planning conflicts early. Commanders can participate more readily in COA analysis even when away from the HQ, with the potential to select a COA without the traditional sequential briefing process. The staff can post plans and orders on interactive web pages or portals for immediate use by subordinate elements (e.g., as facilitated by automated machine-to-machine interfaces or "publish and subscribe" mechanisms). Collaboration capabilities require effective and efficient processes, trained and disciplined users, and a usable collaborative tool infrastructure.

(d) **Information and Intelligence Sharing.** Sharing of information and intelligence with relevant USG departments and agencies, foreign governments and security forces, interorganizational participants, NGOs, and members of the private sector is vital to national security. Commanders at all levels should determine and provide guidance on what information and intelligence needs to be shared with whom and when. DOD information should be secured, shared, and made available throughout the information life cycle to the maximum extent allowed by US laws and DOD policy for foreign disclosure. Commanders, along with their staffs, should share information from the outset of complex operations.

For additional guidance on collaboration and related capabilities, refer to JP 6-0, Joint Communications *System. For additional information on intelligence sharing, refer to JP 2-0,* Joint Intelligence, *and JP 2-01,* Joint and National Intelligence Support to Military Operations.

j. **CCS**

(1) **CCS entails focused efforts to create, strengthen, or preserve conditions favorable for the advancement of national interests, policies, and objectives by understanding and communicating with key audiences through the use of coordinated information, themes, messages, plans, programs, products and actions, synchronized**

Chapter III

> **COMMON OPERATING PRECEPT**
>
> **Inform domestic audiences and shape the perception and attitudes of key foreign audiences as an explicit and continuous operational requirement.**

with the other instruments of national power. Commanders must address and mitigate real or perceived differences between actions and words (the "say-do" gap), since this divergence can reduce our credibility and negatively affect planned operations. Synchronization of IRCs, PA, and other actions is essential for successful CCS.

(2) **Integral to joint planning,** CCDRs and subordinate JFCs should ensure their CCS efforts support higher-level CCS plans, programs, and actions aimed at key audiences. One approach to accomplish this is to develop a CCS LOE and related plans that provide intent, objectives, thematic guidance, and the process to coordinate and integrate CCS-related ways and means. This approach can ensure consistency of messages, activities, and operations to the lowest level with supporting commands, interagency partners, and other relevant stakeholders.

(3) The JFC and staff should include the CCS approach or LOE as part of the commander's intent and operational approach. Plans and orders provide additional guidance and tasks to synchronize the JFC's primary supporting capabilities and actions of PA, IO, and DSPD. CCS products should include a **narrative, themes, and messages, as well as visual products, supporting activities, and identification of key audiences.** These products and activities help guide and regulate joint force actions when communicating and interacting with the local populace, interorganizational participants, and the media, and ensure joint force actions support, align with, and complement other relevant objectives.

(a) The Narrative. A narrative is a short story used to underpin operations and to provide greater understanding and context to an operation or situation. For every military operation, the President or NSC staff may create the national/strategic narrative to explain events in terms consistent with national policy. This guidance should be passed along to military planners and provided to the JFC in the terms of operational orders or other strategic guidance. The end result should be a military plan that aligns both operations and communications with the national strategy and is consistent with the national narrative.

(b) Themes. Strategic themes are developed by the NSC staff, DOS, DOD, and other USG departments and agencies. JFCs support strategic themes by developing operational-level themes appropriate to their mission and authority. Themes at each level of command should support the themes of the next higher level, while also supporting USG strategic themes to ensure consistent communication to local and international.

(c) Messages. Messages support themes by delivering tailored information to a specific public and can also be tailored for delivery at a specific time, place, and communication method. While messages are more dynamic, they must always support the more enduring themes up and down the chain of command. The more dynamic nature and leeway inherent in messages provide joint force communicators and planners more agility in reaching key audiences.

For additional guidance on PA and IO support to CCS, refer to JP 3-61, Public Affairs, *and JP 3-13,* Information Operations. *Also see JP 5-0,* Joint Planning, *for information on communication synchronization planning. For more information on CCS, see JDN 2-13,* Commander's Communication Synchronization.

k. **KLE.** Most operations require commanders and other leaders to engage key local and regional leaders to affect their attitudes and gain their support. **Building relationships to the point of effective engagement and influence usually takes time.** Commanders can be challenged to identify key leaders, develop messages, establish dialogue, and determine other ways and means of delivery, especially in societies where interpersonal relationships are paramount. Commanders use CCS processes to manage messages, delivery, and impacts to ensure deconfliction of messaging efforts. Interaction opportunities with friendly and neutral leaders could include face-to-face meetings, town meetings, and community events. Understanding cultural context, cognitive orientation patterns, and communication methods is essential. The J-2's joint intelligence preparation of the operational environment (JIPOE) should identify key enemy and neutral leaders, as well as key friendly leaders who are not in the commander's sphere of influence. However, the entire staff should identify leaders relative to their functional areas as part of JIPOE.

l. **Risk Management**

(1) Risk management is the process to identify, assess, and control hazards arising from operational factors and make decisions that balance risk cost with mission benefits. It assists organizations and individuals in making informed decisions to reduce or offset risk, thereby increasing operational effectiveness and the probability of mission success. The commander determines the level of risk that is acceptable with respect to aspects of operations and should state this determination in commander's intent. The operation assessment process provides a common method to identify and review risks during planning and execution. Risk is one of the review deliverables of the overall operation assessment activity. The assessment of risk to mission includes an overall risk to mission analysis (e.g., low, moderate, significant, and high) along multiple criteria (e.g., authorities and permissions; policy; forces, basing, and agreements; resources; capabilities; PN contributions; and other USG support). To assist in risk management, commanders and their staffs may develop or institute a risk management process tailored to their mission or OA. Figure III-3 is a generic model that contains the likely elements of a risk management process.

(2) Risk management is a function of command and a key planning consideration. Risk management helps commanders preserve lives and resources; avoid, eliminate, or mitigate unnecessary risk; identify feasible and effective control measures where specific standards do not exist; and develop valid COAs. Prevention of friendly fire incidents is a key consideration in risk management. However, risk management does not inhibit a commander's flexibility and initiative, remove risk altogether (or support a zero-defects

Chapter III

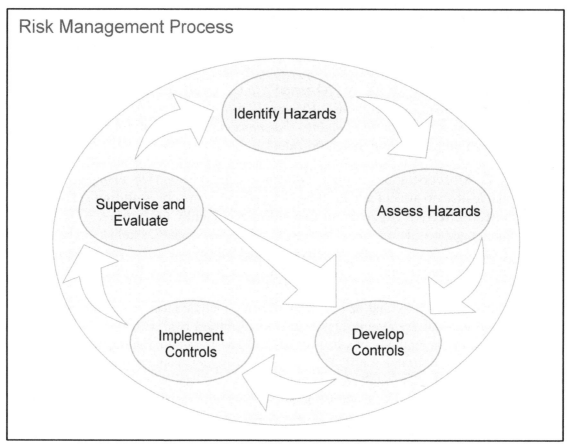

Figure III-3. Risk Management Process

mindset), dictate a go/no-go decision to take a specific action, sanction or justify violating the law, or remove the necessity for SOPs. Risk management is relevant at all levels, across the range of military operations, and through all phases of an operation and its branches and sequels. To alleviate or reduce risk, commanders may take a variety of actions, such as changing the CONOPS, changing the plan for employment of fires, executing a branch to the original plan, or employing countermeasures.

(3) **Safety** preserves military power. High-tempo operations may increase the risk of injury and death due to mishaps. Command interest, discipline, risk mitigation measures, education, and training lessen those risks. The JFC reduces the chance of mishap by conducting risk assessments, assigning a safety officer and staff, implementing a safety program, and seeking advice from local personnel. Safety planning factors could include geospatial and weather data, local road conditions and driving habits, identifying uncharted or uncleared mine fields, and special equipment hazards.

m. **PA.** PA are communication activities with external and internal audiences. Joint PA plans, coordinates, and synchronizes US military public information activities and resources to support the CCS and operational objectives through the distribution of truthful, timely, and factual information about joint military activities. PA contributes to the achievement of military objectives by countering incorrect information and adversary and enemy propaganda through the dissemination of accurate information. PA also observes

Joint Functions

operations security (OPSEC) procedures by educating the media on the implications of premature release of information or the inadvertent release of classified or sensitive information identified in the CCIRs, along with managing the appropriate release of information to the public. PA advises the JFCs on the possible impact of military operations and activities on public perception.

(1) The speed and methods with which people and organizations can collect and convey information makes it probable that incidents will be quickly publicized worldwide. Internet sites, social media, text messages, and mobile smartphones are some of the means through which adversaries communicate. This instantaneous, unfiltered and often incomplete, intentionally biased, or factually incorrect information provided via satellite and the Internet makes planning and effective execution of PA essential.

(2) PA support is important in every phase of operations. The JFC should develop a communication plan for internal and external audiences that addresses both current and future operations. An effective plan provides proactive ways to communicate information about an operation and fulfills the US military's obligation to keep the public informed while maintaining essential secrecy through OPSEC compliance. This plan should minimize adverse effects upon the joint operation from inaccurate news media reporting and analysis, failure to comply with CCMD OPSEC guidance, and promulgation of disinformation and misinformation. PA support is important in every phase of operations. Communication plans should provide for open, independent reporting, and anticipate and respond to media queries. These plans should provide the maximum disclosure allowed with minimum delay and create an environment between the JFC and the news media that encourages balanced, contextual coverage of operations. An effective plan provides proactive ways to communicate information about an operation and fulfills the US military's obligation to keep the US public informed while maintaining essential secrecy through OPSEC compliance.

For additional guidance on PA, refer to JP 3-61, Public Affairs. *For additional guidance on IO, refer to JP 3-13,* Information Operations.

n. **CMO.** CMO are activities that establish, maintain, influence, or exploit relationships between foreign military forces, indigenous populations, and institutions with the objective to reestablish or maintain stability in a region or HN. During all military operations, CMO can coordinate the integration of military and nonmilitary instruments of national power. CA forces support CMO by conducting military engagement, and humanitarian and civic assistance to influence the populations of the HN and other foreign nations in the OA. CA specializes in indirect approaches to accomplish objectives during both traditional and IW.

For additional guidance on CMO, refer to JP 3-57, Civil-Military Operations.

o. **Military Deception (MILDEC).** Commanders conduct MILDEC to mislead enemy decision makers and commanders and cause them to take specific actions or inactions. The intent is to cause enemy commanders to form inaccurate impressions about friendly force dispositions, capabilities, vulnerabilities, and intentions; misuse their intelligence collection assets; and fail to employ their combat or support units to best advantage. As executed by JFCs, MILDEC targets enemy leaders and decision makers through the manipulation of their intelligence collection, analysis, and dissemination systems. MILDEC depends on intelligence to identify deception targets, assist in developing credible stories, identify and orient on appropriate receivers (the readers of the story), and assess the effectiveness of the deception effort. Deception requires a thorough knowledge of the enemy and their decision-making processes. During the formulation of the CONOPS, planners determine how the JFC wants the enemy to act at critical points in the operation. Those desired enemy actions then become the MILDEC objective. MILDEC is focused on causing the enemy to act in a desired manner, not simply to be misled in their thinking.

MILITARY DECEPTION IN THE YOM KIPPUR WAR, 1973

On 6 October 1973, the Egyptian 3rd Army surprised the Israeli Defense Force by attacking across the Suez Canal. Egyptian forces gained a significant foothold in the Sinai and began to drive deeper until a determined defense and counterattack drove them back.

To achieve the initial surprise, Egyptian forces conducted deception operations of strategic, operational, and tactical significance to exploit Israeli weaknesses. At the strategic level, they conveyed the notions that they would not attack without both a concerted Arab effort and an ability to neutralize the Israeli Air Force, and that tactical preparations were merely in response to feared Israeli retaliation for Arab terrorist activity. At the operational level, Egyptian forces portrayed their mobilization, force buildup, and maneuvers as part of their annual exercises. Egyptian exercises portraying an intent to cross the canal were repeated until the Israelis became conditioned to them and therefore did not react when the actual attack occurred. At the tactical level, Egyptian forces expertly camouflaged their equipment, denying information to Israeli observers and creating a false impression of the purpose of the increased activity.

For their part, Israeli forces were overconfident and indecisive at the operational and strategic levels. In spite of the deception, tactical observers reported with increasing urgency that the Egyptian buildup and activity were significant. Their reports caused concern, but no action. Egyptian forces exploited these vulnerabilities and timed the attack to occur on Yom Kippur, the Jewish Day of Atonement, when they perceived the response of Israeli forces would be reduced. The Israeli Defense Force intelligence convinced itself that the Arabs would be unable to attack for a number of years, and they ignored warnings intelligence.

As a result of their deception efforts, synchronized with other operations of the force, Egyptian forces quickly and decisively overwhelmed Israeli forces in the early stages of the Yom Kippur War.

Various Sources

For additional guidance on MILDEC, refer to JP 3-13.4, Military Deception.

p. **Language, Regional, and Cultural Expertise.** Language skills, regional knowledge, and cultural awareness enable effective joint operations. Deployed joint forces should understand and effectively communicate with HN populations, local and national government officials, and multinational partners. This capability is best built on analysis of national, regional, and local culture, economy, politics, religion, and customs. Lessons learned from Operation IRAQI FREEDOM (OIF) and Operation ENDURING FREEDOM (OEF) indicate these capabilities can save lives and are integral to mission accomplishment. Consequently, commanders should integrate training and capabilities for foreign language and regional expertise capabilities in contingency, campaign, and supporting plans, and provide for them in support of daily operations and activities.

For specific planning guidance and procedures regarding language and regional expertise, refer to CJCSI 3126.01, Language, Regional Expertise, and Culture (LREC) Capability Identification, Planning, and Sourcing.

For additional and more detailed guidance on C2 of joint forces, refer to JP 1, Doctrine for the Armed Forces of the United States.

For additional guidance on C2 of air, land, or maritime operations, refer to JP 3-30, Command and Control of Joint Air Operations; *JP 3-31,* Command and Control for Joint Land Operations; *and JP 3-32,* Command and Control for Joint Maritime Operations.

For additional guidance on C2 of SOF, refer to JP 3-05, Special Operations.

3. **Intelligence**

a. **Understanding the OE is fundamental to joint operations.** The intelligence function supports this understanding with analysis of the OE to inform JFCs about adversary capabilities, COGs, vulnerabilities, and future COAs and to help commanders and staffs understand and map friendly, neutral, and threat networks. Using the continuous JIPOE analysis process, properly tailored JIPOE products can enhance OE understanding and enable the JFC to act inside the enemy's decision cycle. Intelligence activities and assessments also occur while defending the homeland within the guidelines of applicable regulations and laws.

b. Tailored continuous JIPOE products support JPP steps 2-7 and the four planning functions starting with an OE baseline characterization to facilitate planning. Throughout execution, tailored continuous JIPOE products capture the dynamic OE in support of the assessment process to facilitate risk management and operations adjustments, and to identify new opportunities. Because joint forces can suffer casualties due to various health threats such as disease and combat or noncombat injuries, medical intelligence preparation of the operational environment (MIPOE) products help decision makers devise protection measures to mitigate these threats.

For further information on MIPOE, refer to JP 4-02, Joint Health Services.

Chapter III

c. People and organizations other than the enemy may positively or negatively affect the friendly mission. These actors may include the population, HN government, and potential opposition leaders. Other relevant actors may include international organizations, non-state actors, and NGOs. By first identifying the relevant actors and learning as much as possible about them and their interrelationships, the JFC can develop an approach that will facilitate decision making and behavior (active or passive) among relevant actors that is consistent with the desired end state of the operation. Sociocultural analysis and identity intelligence (I2) activities enable a better understanding of the relevant actors. Individuals may fit into more than one category of actor. For example, a tribal leader may also work as a district governor, while also working behind the scenes to provide financial and material support to an insurgency.

d. **JIPOE**

(1) JFCs use assigned and attached intelligence forces and coordinate supporting interagency intelligence capabilities to develop a current intelligence picture and analyze the OE. These supporting capabilities include combat support agencies (e.g., National Security Agency, Defense Intelligence Agency, and National Geospatial-Intelligence Agency [NGA]) and national intelligence agencies (e.g., Central Intelligence Agency). National intelligence support may be provided to the J-2 as requested to integrate national intelligence capabilities into a comprehensive intelligence effort designed to support the joint force. J-2 should integrate these supporting capabilities with the efforts of the assigned and attached intelligence forces. Liaison personnel from the various agencies provide access to the entire range of capabilities resident in their agencies and can focus those capabilities on the JFC's intelligence requirements.

(2) As crises emerge that potentially require military action, JFCs examine available intelligence estimates. As part of the JIPOE process, JFCs focus intelligence efforts to determine or confirm enemy COGs and refine estimates of enemy capabilities, dispositions, intentions, and probable COAs within the context of the current situation. They look for specific warning intelligence of imminent enemy activity that may require an immediate response or an acceleration of friendly decision cycles.

For additional guidance on intelligence support to JIPOE, refer to JP 2-01.3, Joint Intelligence Preparation of the Operational Environment.

e. The **intelligence function** encompasses the joint intelligence process. The joint intelligence process consists of six interrelated categories of intelligence operations:

(1) Planning and direction of intelligence activities.

(2) Collection of data.

(3) Processing and exploitation of collected data to produce relevant information.

(4) Analysis of information and production of intelligence.

(5) Dissemination and integration of intelligence with operations.

(6) Evaluation and feedback regarding intelligence effectiveness and quality.

f. **Key Considerations**

(1) **Responsibilities. JFCs and their component commanders are the key players in planning and conducting intelligence tasks.** Commanders are more than just consumers of intelligence. They are responsible for fully integrating intelligence into their plans and operations. They are also responsible for distributing intelligence and information to subordinate commands, and when appropriate, to relevant participants through established protocols and systems. Commanders establish operational and intelligence requirements and continuous feedback to ensure optimum intelligence support to planning and operations. This interface supports the commander and operational planning and execution. It also mitigates surprise, assists friendly deception efforts, and enables joint operation assessment.

(2) **Surveillance and Reconnaissance.** Surveillance and reconnaissance support information collection across the OA. These activities focus on planned collection requirements, but are also sufficiently flexible to respond to time-sensitive and emerging requirements. Commanders will also require persistent surveillance of specific targets that are mission essential and support guidance and intent. Cyberspace ISR includes activities in cyberspace conducted under an execute order to gather intelligence to support future operations. Cyberspace ISR focuses on tactical and operational intelligence and on mapping enemy and adversary networks to support military planning. Cyberspace ISR requires appropriate deconfliction and cyberspace forces that are trained and certified to a common standard with the intelligence community.

(3) **CI and Human Intelligence (HUMINT).** JFCs rely on intelligence acquired through CI and HUMINT capabilities throughout all phases of joint operations. CI and HUMINT both use human sources to collect information, and while their activities may at times be overlapping, each has its own distinct purpose and function. CI support is used to protect the force and must be fully integrated into planning and execution. CI collection gathers information that will assist in identifying, exploiting, and neutralizing threats posed by terrorists and foreign intelligence entities. HUMINT collection strives to satisfy commander's PIRs. CI attempts to deny or shape the enemy's or adversary's knowledge of the US while HUMINT shapes the joint force's understanding of the enemy or adversary.

(4) **I2.** I2 is gathered from identity attributes of individuals, groups, networks, or populations of interest. Regional and global trends have placed greater requirements on the JFC to be able to recognize and differentiate one person from another to support protection and intelligence functions. I2 activities help the joint force protect and control relevant populations. I2 products, such as **biometric watch lists** and **persons of interest overlays,** assist US forces, the HN, and PNs to positively identify, track, characterize, and disrupt threat actors. These products enhance understanding of how these actors and networks can affect the OE.

Chapter III

For additional information on CI and I2, refer to JP 2-01.2, Counterintelligence and Human Intelligence in Joint Operations.

For additional guidance on the intelligence function, refer to JP 2-0, Joint Intelligence; *JP 2-01,* Joint and National Intelligence Support to Military Operations; *and other subordinate JPs that address intelligence support to targeting, CI, HUMINT, geospatial intelligence (GEOINT), and JIPOE.*

4. Fires

a. To **employ fires** is to use available weapons and other systems to create a specific effect on a target. Joint fires are those delivered during the employment of forces from two or more components in coordinated action to produce desired results in support of a common objective. Fires typically produce destructive effects, but various other ways and means can be employed with little or no associated physical destruction. This function encompasses the fires associated with a number of tasks, missions, and processes, including:

(1) **Conduct Joint Targeting.** This is the process of selecting and prioritizing targets and matching the appropriate response to them, taking account of command objectives, operational requirements, and capabilities.

(2) **Provide Joint Fire Support.** This task includes joint fires that assist joint forces to move, maneuver, and control territory, populations, space, cyberspace, airspace, and key waters.

(3) **Countering Air and Missile Threats.** This task integrates offensive and defensive operations and capabilities to attain and maintain a desired degree of air superiority and force protection. These operations are planned to destroy or negate enemy manned and unmanned aircraft and missiles, both before and after launch.

(4) **Interdict Enemy Capabilities.** Interdiction diverts, disrupts, delays, or destroys the enemy's military surface capabilities before they can be used effectively against friendly forces, or to otherwise achieve their objectives.

(5) **Conduct Strategic Attack.** This task includes offensive action against targets—whether military, political, economic, or other—which are selected specifically in order to achieve national or military strategic objectives.

(6) **Employ IRCs.** IRCs are tools, techniques, or activities employed within the information environment to create effects and operationally desirable conditions. In the context of the fires function, this task focuses on the integrated employment of IRCs in concert with other LOOs and LOEs, to influence, disrupt, corrupt, or usurp an enemy's decision making.

(7) **Assess the Results of Employing Fires.** This task includes assessing the effectiveness and performance of fires as well as their contribution to the larger operation or objective.

Joint Functions

b. **Key Considerations.** The following are key considerations associated with the above tasks.

(1) **Targeting** supports the process of linking the desired effects of fires to actions and tasks at the component level. Commanders and their staffs must consider strategic and operational-level objectives, the potential for friendly fire incidents and other undesired fires effects, and operational limitations (e.g., constraints and restraints) when making targeting decisions. Impact on all systems in the OE should be considered during this process. Successful integration of nonlethal engagement options through cyberspace into the targeting process is important to mission accomplishment in many operations.

(a) **Oversight.** JFCs may task their staff to accomplish broad targeting oversight functions or may delegate the responsibility to a subordinate commander. **Typically, JFCs organize joint targeting coordination boards (JTCBs).** If the JFC so designates, a JTCB may be either an integrating center for this effort or a JFC-level review mechanism. In either case, it should be composed of representatives from the staff, all components and, if required, their subordinate units. The primary focus of the JTCB is to ensure target priorities, guidance, and the associated desired results link to the JFC's objectives. Briefings conducted at the JTCB should ensure that all components and applicable staff elements coordinate and synchronize targeting efforts with intelligence and operations.

(b) **Delegation of Joint Targeting Process Authority.** The JFC is responsible for all aspects of the targeting process and may conduct joint targeting at the joint force HQ level or authorize a component commander to do so. The JFC normally appoints the deputy JFC or a component commander to chair the JTCB. When the JFC does not delegate targeting authority and does not establish a JTCB, the JFC performs this task at the joint force HQ with the assistance of the J-3. In this instance, the JFC may approve the formation within the J-3 of a joint fires element to provide recommendations to the J-3. The JFC ensures this process is a joint effort involving applicable subordinate commands. Whoever the JFC authorizes to plan, coordinate, and deconflict joint targeting must possess or have access to a sufficient C2 infrastructure, adequate facilities, joint planning expertise, and appropriate intelligence.

For additional targeting guidance, refer to JP 3-60, Joint Targeting.

(c) **Air Apportionment.** In the context of joint fires, air apportionment is part of the targeting process. Air apportionment is how the JFACC ensures the weight of joint force air effort is consistent with the JFC's intent and objectives. After consulting with other component commanders, the JFACC recommends air apportionment to the JFC, who makes the air apportionment decision. The JFACC's rationale for the recommendation may include priority or percentage of effort based on the JFC's CONOPS, specific tasks for which air assets are essential, and other factors such as the component commanders' joint fires requirements. Following the JFC's air apportionment decision, the JFACC allocates and tasks the capabilities/forces made available.

Chapter III

For additional guidance on air apportionment, refer to JP 3-30, Command and Control of Joint Air Operations.

(2) **Joint fire support** includes joint fires that assist air, land, maritime, cyberspace, and special operations forces to move, maneuver, and control territory, populations, airspace, cyberspace, EMS, and key waters. Joint fire support may include, but is not limited to, manned and unmanned fixed-wing, rotary-wing, and tiltrotor aircraft capabilities; naval surface fire support; artillery, mortars, rockets, and missiles; and other effects of some cyberspace attack, space control operations, EA, and other nonlethal capabilities. Close air support is a critical element of joint fire support that requires detailed planning, coordination, and training of ground and supporting air forces for safe and effective execution. Integration and synchronization of joint fires and joint fire support with the fire and maneuver of the supported force are essential.

For additional guidance on joint fire support, refer to JP 3-09, Joint Fire Support. *For more information on close air support, see JP 3-09.3,* Close Air Support.

(3) **Countering Air and Missile Threats**

(a) The JFC counters air and missile threats to ensure friendly freedom of action, provide protection, and deny enemy freedom of action. Counterair integrates offensive and defensive operations to attain and maintain a desired degree of air superiority and protection by neutralizing or destroying enemy aircraft and missiles, both before and after launch. The counterair mission is inherently a joint and interdependent endeavor. Each component of the joint force contributes capabilities necessary for mission success. In addition, Service capability and force structure development reflect a purposeful reliance on all components to maximize complementary and reinforcing effects while minimizing relative vulnerabilities. Due to the joint and interdependent nature, all components of the joint force normally are tasked to conduct operations in support of the counterair mission. The JFC will normally designate an AADC and a JFACC to enhance unity of command (or unity of effort), centralized planning and direction, and decentralized execution for countering air and missile threats.

(b) **Offensive counterair (OCA)** operations are the preferred method of countering theater air and missile threats. OCA typically seeks to dominate enemy airspace and destroy, disrupt, or neutralize enemy aircraft, missiles, launch platforms, and their supporting structures as close to their sources as possible before and after launch. OCA includes attack operations, fighter sweep, fighter escort, and suppression of enemy air defenses. DCA normally attempts to degrade, neutralize, or defeat enemy air and missile attacks attempting to penetrate friendly airspace. Both OCA and DCA may also be conducted to ensure access and freedom of action in international airspace. These operations may use aircraft, surface-to-surface missiles, surface-to-air missiles, artillery, ground forces, special operations, cyberspace attack, and EA. US forces must be capable of countering the air and missile threats during all phases of an operation.

(c) **Control of the Air.** Control of the air is a prerequisite to success for modern operations or campaigns because it prevents enemy air and missile threats from

Joint Functions

effectively interfering with operations thus facilitating freedom of action and movement. Control of the air cannot be assumed. In the air, the degree of control can range from no control, to parity where neither opponent can claim any level of control over the other, to local air superiority to air supremacy over the entire OA. Control of the air may vary over time. It is important to remember that the degree of control of the air is scalable and can be localized geographically (horizontally and vertically) or defined in the context of an entire theater. The desired degree of control will be at the direction of the JFC and based on the JFC's CONOPS, and will typically be an initial priority objective of joint air operations. Air superiority is that degree of control of the air by one force that permits the conduct of its operations at a given time and place without prohibitive interference from air and missile threats. Air supremacy is that degree of control of the air wherein the opposing force is incapable of effective interference within the OA using air and missile threats. Counterair operations usually begin early in the conduct of an operation or campaign to produce the desired degree of control of the air at the times and places chosen by the JFC.

(d) **Integrating Air and Missile Defense.** While joint combat focuses on operations within one or more OAs, threats to joint forces can come from well outside assigned JOAs, and even outside a GCC's AOR. In particular, an enemy's ballistic and cruise missiles and long-range aircraft can pose significant challenges that require integration of defensive capabilities from both within and beyond a GCC's AOR. The GCC integrates air and missile defense capabilities and activities within the theater. In support, SecDef establishes command relationships for global missile defense, global strike, and other cross-AOR operations. CDRUSSTRATCOM synchronizes planning for global missile defense. The intended result is integration of OCA attack operations, DCA operations, and other capabilities as required to create the JFC's desired effects.

Refer to JP 3-01, Countering Air and Missile Threats, *for additional guidance on air superiority and countering air and missile threats.*

(4) **Interdiction**

(a) Interdiction is a powerful tool for JFCs. Interdiction operations are actions to divert, disrupt, delay, or destroy the enemy's military surface capability before it can be used effectively against friendly forces to achieve enemy objectives. Air interdiction is conducted at such distance from friendly forces that detailed integration of each air mission with the fire and movement of friendly forces is not required. The JFC plans and synchronizes the overall interdiction effort in the assigned JOA. The JFACC is normally the supported commander for the JFC's overall air interdiction effort, while JFLCCs and JFMCCs are supported commanders for interdiction in their AOs.

(b) Interdiction can include operations conducted under lawful authority to track, identify, divert, delay, intercept, board, detain or destroy vessels, vehicles, aircraft, people, cargo, and money. At the direction of appropriate authorities, forces conducting HD or DSCA may perform interdiction against specific targets. For example, military forces also provide DSCA to USG departments or agencies responsible for domestic law enforcement interdiction activities when requested and approved by SecDef or the

President. These activities include actions to divert, disrupt, delay, intercept, board, detain, or destroy, as appropriate, suspect vessels, vehicles, aircraft, people, and cargo. Counterproliferation interdiction is a proactive USG mission intended to deter, make more costly, inhibit, disrupt, and prevent trafficking in shipments of proliferation concern. All such DOD operations and activities shall be consistent with US domestic law and international law, applicable DOD policy and SecDef-approved execution orders. Federal law and DOD policy impose limitations on the types of support that the US military may provide and what type of military mission (e.g., HD or DSCA) is being conducted.

See Department of Defense Instruction (DODI) 3025.21, Defense Support of Law Enforcement Agencies, *and DODD 3025.18,* Defense Support of Civil Authorities (DSCA), *for more information on DSCA regarding domestic law enforcement interdiction activities.*

(c) Many elements of the joint force can conduct interdiction operations. Air, land, maritime, space, cyberspace, and special operations forces can conduct interdiction operations as part of their larger or overall mission. For example, naval expeditionary forces charged with seizing and securing a lodgment along a coast may include the interdiction of opposing land and maritime forces inside the amphibious objective area (AOA) as part of the overall amphibious plan.

(d) JFCs may choose to employ interdiction as a principal means to achieve an objective (with other components supporting the component leading the interdiction effort). For example, one of the JFC's objectives associated with the seize initiative phase of an operation might be to prevent the enemy's navy from interfering with friendly force sea transit through a choke point in the JOA. The JFC might task the JFACC to accomplish this through an interdiction effort supported by SOF. Interdiction during warfighting is not limited to any particular region of the OA, but generally is conducted forward of or at a distance from friendly surface forces. Likewise, military interdiction that supports HD is guided and restricted by domestic law to a greater extent than other interdiction. Joint interdiction can be planned to create tactical, operational, or strategic advantages for the joint force, with corresponding adverse effects on the enemy. Interdiction deep in the enemy's rear area can have broad operational effects; however, deep interdiction may have a delayed effect on land, maritime, and selected special operations. Interdiction closer to joint forces will have more immediate operational and tactical effects. Thus, JFCs vary the emphasis upon interdiction operations and surface maneuvers, depending on the situation confronting them.

(e) Counter threat finance (CTF) incorporates efforts to interdict money that funds terrorism, illegal narcotics networks, weapons proliferation, espionage, and other activities that generate revenue through trafficking networks. Illicit finance networks are a critical vulnerability of state and non-state adversaries threatening US national security. Employing CTF activities are the means to detect, counter, contain, disrupt, deter, or dismantle these illicit financial networks. Monitoring, assessing, analyzing, and exploiting financial information are key support functions for CTF activities.

Joint Functions

> **AIR INTERDICTION DURING OPERATION IRAQI FREEDOM**
>
> During Operation IRAQI FREEDOM, most of the effort against Iraqi ground troops was focused on Republican Guard divisions and on a handful of stalwart regular divisions that formed part of the defensive ring south of Baghdad.
>
> One prominent air interdiction success story involved the Iraqi Republican Guard's redeployment of elements of the Hammurabi, Nebuchadnezzar, and Al Nida divisions after 25 March 2003 to the south of Baghdad toward Karbala, Hillah, and Al Kut. Their road movements were steadily bombed by US Air Force A-10s and B-52s (dropping 500-pound bombs) and British Tornados. An Iraqi commander concluded that their movement south had been one of the Iraqi regime's major errors because it exposed the Republican Guard to coalition air power and resulted in large casualty figures.
>
> **SOURCE: Project on Defense Alternatives Briefing Memo #30**
> **Carl Conetta, 26 September 2003**

Refer to JP 3-03, Joint Interdiction, *for more guidance on joint interdiction operations. Refer to JP 3-25,* Countering Threat Networks, *for more information on counter threat finance.*

(5) **Strategic Attack.** A strategic attack is a JFC-directed offensive action against a target—whether military or other—that is selected to achieve national or military strategic objectives. These attacks seek to weaken the enemy's ability or will to engage in conflict or continue an action and as such, could be part of a campaign or major operation, or conducted independently as directed by the President or SecDef. Additionally, these attacks may achieve strategic objectives without necessarily having to achieve operational objectives as a precondition. Suitable targets may include but are not limited to enemy strategic COGs. All components of a joint force may have capabilities to conduct strategic attacks.

(6) **Global Strike**

(a) Global strike is the capability to rapidly plan and deliver extended-range attacks, limited in duration and scope, to create precision effects against enemy assets in support of national and theater commander objectives. Global strike missions employ lethal and non-lethal capabilities against a wide variety of targets.

(b) The UCP assigns CDRUSSTRATCOM the responsibility for global strike. CDRUSTRATCOM plans global strike in full partnership with affected CCDRs. The CJCS or SecDef determines supporting and supported command relationships for execution. In some circumstances, USSTRATCOM may act in a supporting role to a supported CCMD for both global strike planning and execution.

Chapter III

(7) **Limiting Collateral Damage.** Collateral damage is unintentional or incidental injury or damage to persons or objects that would not be lawful military targets based on the circumstances existing at the time. Causing collateral damage does not violate the law of war so long as the collateral damage caused is not excessive in relation to the concrete and direct military advantage anticipated from the attack. Under the law of war, the balancing of military necessity in relation to collateral damage is known as the principle of **proportionality.** Moreover, limiting collateral damage is often an operational or strategic imperative. Further, limiting collateral damage will not only reduce the requirement to address civilian claims but may help better support friendly and HN actions to influence the population and reduce the magnitude and duration of stability activities. Collateral damage assessment is conducted during targeting, and is especially important when considering strikes against WMD storage and production targets.

(8) **Nonlethal Capabilities.** Nonlethal capabilities can generate effects that limit collateral damage, reduce risk to civilians, and may reduce opportunities for enemy or adversary propaganda. They may also reduce the number of casualties associated with excessive use of force, limit reconstruction costs, and maintain the good will of the local populace. Some capabilities are nonlethal by design, and include, but are not limited to, blunt impact and warning munitions, acoustic and optical warning devices, and vehicle and vessel stopping systems.

(a) **Cyberspace Attack.** Cyberspace attack actions create various direct denial effects in cyberspace (i.e., degradation, disruption, or destruction) and manipulation that leads to denial that is hidden or that manifests in the physical domains.

(b) **EA.** EA involves the use of electromagnetic energy, directed energy, or anti-radiation weapons to attack personnel, facilities, or equipment to degrade, neutralize, or destroy enemy combat capability. EA can be used against a computer when the attack occurs through the EMS. Cyberspace attacks use the data stream through the network to deliver logical payloads that create desired effects such as degradation, disruption, and destruction. Integration and synchronization of EA with maneuver, C2, and other joint fires are essential. EW is a component of JEMSO used to exploit, attack, protect, and manage the EME to achieve commander's objectives. EW can be used as a primary capability or used to facilitate delivery of other IRCs through the targeting process.

For additional guidance on cyberspace attack, refer to JP 3-12, Cyberspace Operations. *For additional guidance on EA, refer to JP 3-13.1,* Electronic Warfare. *For additional guidance on JEMSO, refer to CJCSI 3320.01,* Joint Electromagnetic Spectrum Operations (JEMSO).

(c) **Military Information Support Operations (MISO).** MISO convey selected information and indicators to foreign audiences to influence their emotions, motives, and objective reasoning, and ultimately induce or reinforce foreign attitudes and behavior favorable to the originator's objectives. MISO craft messages using a variety of print, audio, audio-visual, and electronic media, which can then be delivered to target audiences. MISO have strategic, operational, and tactical applications and should be considered early in planning to maximize effectiveness. The JFC should address the approval authorities for MISO during planning to ensure integration and unity of effort between MISO, DSCA, and other IRCs.

For additional guidance on MISO, refer to JP 3-13.2, Military Information Support Operations. *MISO support to non-US military is outlined in DODD S-3321-1,* Overt Psychological Operations Conducted by the Military Services in Peacetime and in Contingencies Short of Declared War (U).

(d) **Nonlethal Weapons.** Nonlethal weapons are weapons, devices, and munitions that are explicitly designed and primarily employed to incapacitate targeted personnel or materiel immediately, while minimizing fatalities, permanent injury to personnel, and undesired damage to property in the target area or environment. Nonlethal weapons are intended to have reversible effects on personnel and materiel. Planners should consider nonlethal weapons to minimize loss of life and damage to property that could negatively influence public perception.

5. **Movement and Maneuver**

a. This function encompasses the disposition of joint forces to conduct operations by securing positional advantages before or during combat operations and by exploiting tactical success to achieve operational and strategic objectives. This function includes moving or deploying forces into an OA and maneuvering them to operational depths for offensive and defensive purposes. It also includes assuring the mobility of friendly forces. The **movement and maneuver function** encompasses a number of tasks including:

(1) Deploy, shift, regroup, or move joint and/or component force formations within the OA by any means or mode (i.e., air, land, or sea).

(2) Maneuver joint forces to achieve a position of advantage over an enemy.

(3) Provide mobility for joint forces to facilitate their movement and maneuver without delays caused by terrain or obstacles.

(4) Delay, channel, or stop movement and maneuver by enemy formations. This includes operations that employ obstacles (i.e., countermobility), enforce sanctions and embargoes, and conduct blockades.

(5) Control significant areas in the OA whose possession or control provides either side an operational advantage.

b. **Movement to Attain Operational Reach**

(1) Forces, sometimes limited to those that are forward-deployed or even multinational forces formed specifically for the task at hand, can be positioned within operational reach of enemy COGs or decisive points to achieve decisive force at the appropriate time and place. Operational reach is the distance and duration across which a joint force can successfully employ its military capabilities. At other times, mobilization and deployment processes can be called up to begin the movement of reinforcing forces

Chapter III

from the continental United States (CONUS) or other theaters to redress any unfavorable balance of forces and to achieve decisive force at the appropriate time and place.

(2) JFCs must carefully consider the movement of forces and whether to recommend the formation and/or movement of multinational forces. They must be aware of A2/AD threats which may slow or disrupt the deployment of friendly forces. At times, movement of forces can contribute to the escalation of tension, while at other times its deterrent effect can reduce those tensions. Movement of forces may deter adversary aggression or movement.

Refer to JP 3-35, Deployment and Redeployment Operations, *for more information on the deployment process.*

c. **Maneuver is the employment of forces in the OA through movement in combination with fires to achieve a position of advantage in respect to the enemy.** Maneuver of forces relative to enemy COGs can be key to the JFC's mission accomplishment. Through maneuver, the JFC can concentrate forces at decisive points to achieve surprise, psychological effects, and physical momentum. Maneuver also may enable or exploit the effects of massed or precision fires.

(1) The principal purpose of maneuver is to place the enemy at a disadvantage through the flexible application of movement and fires. The goal of maneuver is to render opponents incapable of resisting by shattering their morale and physical cohesion (i.e., their ability to fight as an effective, coordinated whole) by moving to a point of advantage to deliver a decisive blow. This may be achieved by attacking enemy forces and controlling territory, airspace, EMS, populations, key waters, and LOCs through air, land, and maritime maneuvers.

(2) There are multiple ways to attain positional advantage. A naval expeditionary force with airpower, cruise missiles, and amphibious assault capability, within operational reach of an enemy's COG, has positional advantage. In like manner, land and air expeditionary forces that are within operational reach of an enemy's COG and have the means and opportunity to strike and maneuver on such a COG also have positional advantage. Maintaining full-spectrum superiority contributes to positional advantage by facilitating freedom of action. See Chapter VIII, "Major Operations and Campaigns," paragraph 5.g, "Full-Spectrum Superiority."

(3) At all levels of warfare, successful maneuver requires not only fire and movement but also agility and versatility of thought, plans, operations, and organizations. It requires designating and then, if necessary, shifting the main effort and applying the principles of mass and economy of force.

(a) At the strategic level, deploying units to and positioning units within an operational area are forms of maneuver if such movements seek to gain positional advantage. Strategic maneuver should place forces in position to begin the phases or major operations of a campaign.

Joint Functions

(b) At the operational level, maneuver is a means by which JFCs set the terms of battle by time and location, decline battle, or exploit existing situations. Operational maneuver usually takes large forces from a base of operations to an area where they are in position of operational reach from which to achieve operational objectives. The enemy may use AD actions to impede friendly operations when A2 actions fail. The objective for operational maneuver is usually a COG or decisive point.

(c) At the tactical level, maneuver is a means by which component commanders employ their forces in combination with fires to achieve positional advantage in respect to the enemy.

(4) Force posture (forces, footprints, and agreements) affects operational reach and is an essential maneuver-related consideration during theater strategy development and adaptive planning. Force posture is the starting position from which planners determine additional contingency basing requirements to support specific contingency plans and crisis action responses. These requirements directly support the development of operational LOCs and LOOs and affect the combat power and other capabilities that a joint force can generate. In particular, the arrangement and positioning of temporary contingency bases underwrite the ability of the joint force to project power by shielding its components from enemy action and protecting critical factors such as sortie or resupply rates. Incomplete planning for contingency base operations can unnecessarily increase the sustainment requirements of the joint force, leading to unanticipated risk. Political and diplomatic considerations can often affect basing decisions. US force basing options span the range from permanently based forces to temporary sea basing that accelerates the deployment and employment of maritime capabilities independent of infrastructure ashore.

(5) JFCs should consider various ways and means to help maneuver forces attain positional advantage. Specifically, combat engineers provide mobility of the force by breaching obstacles, while simultaneously countering the mobility of enemy forces by emplacing obstacles, and minimizing the effects of enemy actions on friendly forces.

6. Protection

a. The protection function encompasses force protection, force health protection (FHP), and other protection activities.

(1) The function focuses on **force protection,** which preserves the joint force's fighting potential in four primary ways. One way uses active defensive measures that protect the joint force, its information, its bases, necessary infrastructure, and LOCs from an enemy attack. Another way uses passive defensive measures that make friendly forces, systems, and facilities difficult to locate, strike, and destroy by reducing the probability of, and minimizing the effects of, damage caused by hostile action without the intention of taking the initiative. The application of technology and procedures to reduce the risk of friendly fire incidents is equally important. Finally, emergency management and response reduce the loss of personnel and capabilities due to isolating events, accidents, health threats, and natural disasters.

Chapter III

(2) Force protection does not include actions to defeat the enemy or protect against accidents, weather, or disease. FHP complements force protection efforts by promoting, improving, preserving, or restoring the mental or physical well-being of Service members.

(3) **As the JFC's mission requires,** the protection function also extends beyond force protection to encompass protection of US noncombatants.

b. The **protection function** encompasses a number of tasks, including:

(1) Provide air, space, and missile defense.

(2) Protect US civilians and contractors authorized to accompany the force.

(3) Conduct defensive countermeasure operations, including MILDEC in support of OPSEC, counterdeception, and counterpropaganda operations.

(4) Conduct OPSEC, cyberspace defense, cybersecurity, defensive EA, and electronic protection activities.

(5) Conduct PR operations.

(6) Establish antiterrorism programs.

(7) Establish capabilities and measures to prevent friendly fire incidents.

(8) Secure and protect combat and logistics forces, bases, JSAs, and LOCs.

(9) Provide physical protection and security for forces and means, to include conducting operations to mitigate the effects of explosive hazards.

(10) Provide chemical, biological, radiological, and nuclear (CBRN) defense.

(11) Mitigate the effects of CBRN incidents through thorough planning, preparation, response, and recovery.

(12) Provide emergency management and response capabilities and services.

(13) Protect the DODIN using cybersecurity and cyberspace defense measures.

(14) Identify and neutralize insider threats.

(15) Conduct identity collection activities. These include security screening and vetting in support of I2.

c. Protection considerations affect planning in every joint operation. **Campaigns and major operations** involve large-scale combat against a capable enemy. These operations typically will require the full range of protection tasks, thereby complicating both planning and execution. Although the OA and joint force may be smaller for a **crisis response or limited contingency operation,** the mission can still be complex and dangerous, thus

Joint Functions

requiring a variety of protection considerations. Permissive environments associated with **military engagement, security cooperation, and deterrence** still require that commanders and their staffs consider protection measures commensurate with potential risks. These risks may include a wide range of threats such as terrorism, criminal enterprises, environmental threats and hazards, and cyberspace threats. Continuous research and access to accurate, detailed information about the OE along with realistic training can enhance protection activities.

d. Force protection includes preventive measures taken to prevent or mitigate enemy and insider threat actions **against DOD personnel** (to include family members and certain contractors), resources, facilities, and critical information. These actions preserve the force's fighting potential so it can be applied at the decisive time and place and incorporate integrated and synchronized offensive and defensive measures that enable the effective employment of the joint force while degrading opportunities for the enemy. Force protection is achieved through the tailored selection and application of multilayered active and passive measures commensurate with the level of risk. Intelligence sources provide information regarding an enemy or adversary's capabilities against personnel and resources, as well as information regarding force protection considerations. Foreign and domestic LEAs can contribute to force protection through the prevention, detection, response, and investigation of crime, and by sharing information on criminal and terrorist organizations. I2 can identify threat networks in the OA, enhancing the commander's ability to establish effective antiterrorism programs, screening and vetting activities, and lasting relationships with the HN and local population.

e. **Key Considerations**

(1) **Security of forces and means** enhances force protection by identifying and reducing friendly vulnerability to hostile acts, influence, or surprise. Security operations protect combat and logistics forces, bases, JSAs, and LOCs. Physical security includes physical measures designed to safeguard personnel; to prevent unauthorized access to equipment, installations, material, and documents; and to safeguard them against espionage, sabotage, damage, and theft. The physical security process determines vulnerabilities to known threats; applies appropriate deterrent, control, and denial safeguarding techniques and measures; and responds to changing conditions. Functions in physical security include facility security, law enforcement, guard and patrol operations, special land and maritime security areas, and other physical security operations like military working dog operations or emergency and disaster response support. Measures include fencing and perimeter stand-off areas, land or maritime force patrols, lighting and sensors, vehicle barriers, blast protection, intrusion detection systems and electronic surveillance, and access control devices and systems. Physical security measures, like any defense, should be overlapping and deployed in depth.

For additional guidance on physical security measures, refer to JP 3-10, Joint Security Operations in Theater.

(2) **DCA.** DCA supports protection using both active and passive air and missile defense measures.

Chapter III

(a) **Active air and missile defense** includes all direct defensive actions taken to destroy, nullify, or reduce the effectiveness of hostile air and missile threats against friendly forces and assets. It includes the use of aircraft, air and missile defense weapons, EW, and other available weapons. Ideally, integration of systems will allow for a defense in depth, with potential for multiple engagements that increase the probability for success.

(b) **Passive air and missile defense** includes all measures, other than active air and missile defense, taken to minimize the effectiveness of hostile air and missile threats against friendly forces and assets. These measures include camouflage, concealment, deception, dispersion, reconstitution, redundancy, detection and warning systems, and the use of protective construction.

(3) **Global Ballistic Missile Defense.** Global ballistic missile defense is the overarching characterization of cumulative planning and coordination for those defensive capabilities designed to neutralize, destroy, or reduce effectiveness of enemy ballistic missile attacks that cross AOR boundaries.

For additional guidance on countering theater air and missile threats, refer to JP 3-01, Countering Air and Missile Threats.

(4) **Defensive use of IRCs** ensures timely, accurate, and relevant information access while denying enemies and adversaries opportunities to exploit friendly information and information systems for their own purposes.

(a) **OPSEC, as an IRC,** denies the adversary the information needed to correctly assess friendly capabilities and intentions. It is also a tool, hampering the adversary's use of its own information systems and processes and providing the necessary support to all IRCs. OPSEC complements the other IRCs and should be integrated into planning. The purpose of OPSEC is to **reduce the vulnerability** of US and multinational forces to successful adversary exploitation of critical information. Unlike security programs that seek to protect classified information and controlled unclassified information, OPSEC **identifies, controls, and protects unclassified critical information** that is associated with specific military operations and activities. OPSEC applies to all activities that prepare, sustain, or employ forces. The OPSEC process subsequently analyzes friendly actions associated with military operations and other activities to:

<u>1.</u> Identify those actions that may be observed by adversary intelligence systems.

<u>2.</u> Determine what specific indications could be collected, analyzed, and interpreted to derive critical information in time to be useful to adversaries.

For additional guidance on OPSEC, refer to CJCSI 3213.01, Joint Operations Security, *and JP 3-13.3,* Operations Security.

(b) **DCO** include passive and active CO to preserve friendly cyberspace capabilities and protect data, networks, and net-centric capabilities by monitoring,

Joint Functions

analyzing, detecting, and responding to unauthorized activity within DOD information systems and computer networks.

(c) **Cybersecurity** encompasses measures that protect computers, electronic communications systems and services, wired communications, and electronic communications by ensuring their availability, integrity, authentication, confidentiality, and nonrepudiation. Cybersecurity incorporates protection, detection, response, restoration, and reaction capabilities and processes to shield and preserve information and information systems. Cybersecurity for the DODIN requires integration of the capabilities of people, operations, and technology. This capability establishes a multilayer and multidimensional protection to ensure survivability and mission accomplishment. Cybersecurity must account for the possibility that access to the DODIN can be gained from outside of DOD's control.

For additional guidance on cybersecurity, refer to DODI 8500.01, Cybersecurity.

(d) **Defensive Use of EW. Electronic protection** is that division of EW involving action taken to protect personnel, facilities, and equipment from any effects of friendly or enemy use of the EMS that degrade, neutralize, or destroy friendly combat capability. Defensive EA activities use the EMS to protect personnel, facilities, capabilities, and equipment. Examples include self-protection and force protection measures such as use of expendables (e.g., chaff and active decoys), jammers, towed decoys, directed energy infrared countermeasure systems, and counter radio-controlled improvised explosive device (IED) systems.

Refer to JP 3-13.1, Electronic Warfare, *for further guidance.*

(5) **PR.** PR missions use military, diplomatic, and civil efforts to recover and reintegrate isolated personnel. There are five PR tasks (report, locate, support, recover, and reintegrate) necessary to achieve a complete and coordinated recovery of US military personnel, DOD civilians, DOD contractors, and others designated by the President or SecDef. JFCs should consider all individual, component, joint, and multinational partner capabilities available when planning and executing PR missions.

Refer to JP 3-50, Personnel Recovery, *for further guidance on PR.*

(6) **CBRN Defense.** Preparation for potential enemy use of CBRN weapons is integral to joint planning. Even when an enemy does not possess CBRN materiel or WMD, access to materials such as radiation sources and toxic industrial materials is a significant planning consideration. Whether a CBRN attack achieves traditional military objectives, it will likely generate adverse strategic, operational, psychological, economic, and political effects. CBRN defense focuses on avoiding CBRN hazards (contamination), protecting individuals and units from CBRN hazards, and decontaminating personnel and materiel to restore operational capability. CBRN defense may also contribute to the deterrence of enemy WMD use through the enhancement of US forces' CBRN survivability. CBRN defense capabilities may also respond to a noncombatant incident or accidental causes such as toxic industrial chemical incident.

Chapter III

For additional guidance on CBRN defense, refer to JP 3-11, Operations in Chemical, Biological, Radiological, and Nuclear Environments; JP 3-40, Countering Weapons of Mass Destruction; and JP 3-41, Chemical, Biological, Radiological, and Nuclear Response.

(7) **Antiterrorism** programs support force protection by establishing defensive measures that reduce the vulnerability of individuals and property to terrorist acts, to include limited response and containment by local military and civilian forces. These programs also include personal security and defensive measures to protect Service members, high-risk personnel, civilian employees, family members, DOD facilities, information, and equipment. Personal security measures consist of commonsense rules for the on- and off-duty conduct of every Service member. They also include employment of dedicated guard forces and use of individual protective equipment (IPE), hardened vehicles, hardened facilities, and duress alarms. Security of high-risk personnel safeguards designated individuals who, by virtue of their rank, assignment, symbolic value, location, or specific threat, are at a greater risk than the general population. Terrorist activity may be discouraged by varying the installation force protection posture through the use of a random antiterrorism measures program, which may include varying patrol routes; staffing guard posts and towers at irregular intervals; and conducting vehicle and vessel inspections, personnel searches, and identification checks on a set but unpredictable pattern. To be effective, these measures should be highly visible to any hostile elements conducting surveillance.

Refer to JP 3-07.2, Antiterrorism, for additional guidance on antiterrorism.

(8) The intent of **combat identification (CID)** is to accurately distinguish enemy objects and forces in the OE from others to support engagement decisions. CID supports force protection and enhances operations by helping minimize friendly fire incidents and collateral damage. CID is not a substitute for any requirement to conduct positive identification of hostile forces, hostile acts, or a demonstration of hostile intent as required by ROE to engage targets.

(a) Depending on operational requirements, CID characterization may be limited to "friend," "enemy/hostile," "neutral," or "unknown." In some situations, additional characterizations may be required including, but not limited to, class, type, nationality, and mission configuration. CID characterizations, when applied with ROE, enable engagement decisions concerning use or prohibition of lethal weapons and nonlethal capabilities.

(b) The staff should develop CID procedures early during planning. These procedures must be consistent with ROE and should not interfere with the ability of a unit or individual to engage enemy forces. When developing the JFC's CID procedures, important considerations include the missions, capabilities, and limitations of all participants.

Refer to JP 3-09, Joint Fire Support, for additional guidance on CID.

Joint Functions

(9) **Critical infrastructure protection** programs support the identification and mitigation of vulnerabilities to defense critical infrastructure, which includes DOD and non-DOD domestic and foreign infrastructures essential to plan, mobilize, deploy, execute, and sustain US military operations on a global basis. Coordination between DOD entities and other USG departments and agencies; state, territorial, tribal, and local governments; the private sector; and equivalent foreign entities, is key in effective protection of critical assets controlled both by DOD and private entities. Vulnerabilities found in defense critical infrastructure shall be remediated and/or mitigated based on risk management decisions made by responsible authorities. These vulnerability mitigation decisions should be made using all available program areas, including antiterrorism, MILDEC, OPSEC, and force protection.

(10) **Counter-Improvised Explosive Device (C-IED) Operations.** C-IED operations are the organization, integration, and synchronization of capabilities and activities to reduce casualties and mitigate damage caused by IEDs. They include measures to neutralize the infrastructure supporting the production and employment of IEDs; the development of tactics, techniques, and procedures to counter the IED threat at the tactical level; and the technical and forensic exploitation of the device to obtain information to support targeting, improve force protection, identify material sourcing, and identify weapon signatures.

For further guidance on C-IED, refer to JP 3-15.1, Counter-Improvised Explosive Device Operations.

(11) **Identify and Neutralize Insider Threats.** Insider threats (sometimes referred to as "green-on-blue" or "inside-the-wire" threats) may include active shooters, bombers, spies, and other threats embedded within or working with US forces. These threats are typically persons with authorized access, who use that access to commit any of a variety of illicit actions against friendly force personnel, materiel, facilities, and information. Countering these threats involves coordinating and sharing information among security, cybersecurity, CI, law enforcement, and other personnel and staffs. Identity activities support the identification of insider threats. The joint force security coordinator establishes procedures to counter insider threats across the joint force.

For further guidance on countering inside threats, see JP 3-10, Joint Security Operations in Theater. *For more information on identity activities, see JDN 2-16,* Identity Activities.

f. **FHP** complements force protection efforts, and includes all measures taken by the JFC and the Military Health System to promote, improve, and conserve the mental and physical well-being of Service members. These capabilities enable a healthy and fit force, prevent injury and illness, and protect the force from health hazards. FHP measures focus on the prevention of illness and injury. The JFC is responsible to allocate adequate capabilities to identify health threats and implement appropriate FHP measures. **Health threats** are a composite of ongoing or potential enemy actions; occupational, environmental, geographical, and meteorological conditions; endemic diseases; and the employment of CBRN weapons that can reduce the effectiveness of military forces. Therefore, a robust **health surveillance system** is essential to FHP measures. Health

Chapter III

surveillance includes identifying the population at risk; identifying and assessing hazardous exposures; employing specific countermeasures to eliminate or mitigate exposures; and monitoring and reporting battle injury, disease, and non-battle injury trends and other health outcomes. Occupational and environmental health surveillance enhances the joint force's ability to limit all categories of injuries including combat and operational stress, exposure to CBRN, and explosive hazards.

Refer to JP 4-02, Joint Health Services, *for further guidance on FHP.*

g. **Protection of Civilians.** Persons who are neither part of nor associated with an armed force or group, nor otherwise engaged in hostilities are classified as civilians and have protected status under the law of war.

(1) It is US policy that members of the DOD components comply with the law of war during all armed conflicts, however such conflicts are characterized, and in all other military operations. This includes taking measures to protect civilians. In addition, the accountability, credibility, and legitimacy of a joint operation, the success of the overarching mission, and the achievement of US strategic objectives depends on the joint forces' ability to minimize harm to civilians in the course of their own operations and, potentially, their ability to mitigate harm arising from the operations of other parties. Strategic objectives often involve strengthening security, stability, and civilian well-being.

(2) Protection of civilians may be the primary purpose of a mission or a supporting task. The protection of civilians from deliberate attack as a strategic or operational imperative is distinct from the legal obligations of US forces to minimize harm to civilians during the conduct of operations. Effective protection of civilians depends on adaptive units, a command climate that emphasizes its importance, and leaders who can make timely and appropriate decisions based on critical situations on the ground. Moreover, joint forces must have in place ROE that prioritize and account for the protection of civilians in the planning process.

(3) Civilian casualty mitigation directly affects the success of the overall mission. Even tactical actions can have strategic and second-order effects. Minimizing and addressing civilian casualty incidents supports strategic imperatives and is also at the heart of the profession of arms. Failure to minimize civilian casualties can undermine national policy objectives and the mission of joint forces, while assisting the enemy. Additionally, civilian casualties can incite increased opposition to joint forces. Focused attention on civilian casualty mitigation can be an important investment to maintain legitimacy and ensure eventual success.

See Chapters V and X of Department of Defense Manual, Law of War, *regarding obligations for protection of civilians.*

7. Sustainment

a. **Sustainment is the provision of logistics and personnel services to maintain operations through mission accomplishment and redeployment of the force.** Sustainment provides the JFC the means to enable freedom of action and endurance and to

Joint Functions

extend operational reach. Sustainment determines the depth to which the joint force can conduct decisive operations, allowing the JFC to seize, retain, and exploit the initiative. The **sustainment function** includes tasks to:

(1) Coordinate the supply of food, operational energy (fuel and other energy requirements), arms, munitions, and equipment.

(2) Provide for maintenance of equipment.

(3) Coordinate and provide support for forces, including field services; personnel services support; health services; mortuary affairs; religious support; postal support; morale, welfare, and recreational support; financial support; and legal services.

(4) Build and maintain contingency bases.

(5) Assess, repair, and maintain infrastructure.

(6) Acquire, manage, and distribute funds.

(7) Provide common-user logistics support to other government agencies, international organizations, NGOs, and other nations.

(8) Establish and coordinate movement services.

(9) Establish large-scale detention compounds and sustain enduring detainee operations.

b. JFCs should identify sustainment capabilities early in planning. Sustainment should be a priority consideration when the timed-phased force and deployment data is built. Sustainment provides JFCs with flexibility to develop branches and sequels and to refocus joint force efforts. Given mission objectives and adversary threats, the ultimate goal is for planners to develop a feasible, supportable, and efficient CONOPS that takes into account the threat and defense of logistical forces. Prior to the development of contingency plans, CCMDs develop a theater logistics analysis, TLO, and distribution plan to provide detailed mobility and distribution analysis to ensure sufficient capacity or planned enhanced capability is available to support the CCMD's TCP.

c. **Logistics** is planning and executing the movement and support of forces. It concerns the integration of strategic, operational, and tactical support efforts within the theater, while scheduling the mobilization and movement of forces and materiel to support the JFC's CONOPS. The relative combat power that military forces can generate against an enemy is constrained by a nation's capability to plan for, gain access to, and deliver forces and materiel to points of application. **Logistics covers the following core functions: supply, maintenance, deployment and distribution, health services, logistic services, engineering, and operational contract support (OCS).** Associated with these functions, logistics includes those aspects of military operations that deal with:

Chapter III

(1) Materiel acquisition, receipt, storage, movement, distribution, maintenance, evacuation, and disposition.

(2) In-transit visibility and asset visibility.

(3) Common-user logistics support to other USG departments and agencies, international organizations, NGOs, and other nations.

(4) Logistic services (food, water, and ice, contingency basing, and hygiene).

(5) OCS (synchronization of contract support for operations and contract management).

(6) Disposal operations.

(7) Engineering support.

(8) Facilities and infrastructure acquisition, construction, maintenance, operation, and disposition.

(9) Infrastructure assessment, repairs, and maintenance.

(10) Detention compounds (establish and sustain large-scale to support enduring detainee operations).

(11) Host-nation support (HNS).

(12) Personnel movement, including patient movement, evacuation, and hospitalization.

d. **Personnel services** are sustainment functions provided to personnel rather than to systems and equipment. Personnel services complement logistics by planning for and coordinating efforts that provide and sustain personnel during joint operations. These services include:

(1) Human resources support.

(2) Religious support.

(3) Financial management.

(4) Legal support.

(5) Morale, welfare, and recreation support.

For further guidance on logistic support, refer to JP 4-0, Joint Logistics. For further guidance on personnel services, refer to JP 1-0, Joint Personnel Support. For further guidance on legal support, refer to JP 1-04, Legal Support to Military Operations. For further guidance on religious affairs, refer to JP 1-05, Religious Affairs in Joint

Operations. *For further guidance on financial management support, refer to JP 1-06, Financial Management Support in Joint Operations.*

e. **Key Considerations**

(1) **Employment of Logistic Forces.** For some operations, logistic forces may be employed in quantities disproportionate to their normal military roles, and in nonstandard tasks. Further, logistic forces may precede other military forces or may be the only forces deployed. Logistic forces may also continue to support other military personnel and civilians after the departure of combat forces. In such cases, they must be familiar with and adhere to applicable status-of-forces agreements and acquisition and cross-servicing agreements to which the US is a party. Given the potential complexity of OEs, logistic forces must be familiar with and adhere to legal, regulatory, and diplomatic/political restraints governing US involvement because of the specialized nature and unique authorities in operations such as disaster relief and humanitarian assistance. Logistic forces, like all other forces, must be capable of self-defense, particularly if they deploy alone or in advance of other military forces.

(2) **Protection.** Logistics forces must be capable of self-defense, particularly if they deploy alone or in advance of other military forces. However, the JFC should view combat and logistics forces as a unit with a seamless mission and objective, and balance the allocation of security resources accordingly in support of the JFC's mission.

(3) **Facilities.** JFCs should plan for the early acquisition (leasing) of real estate and facilities and bases when temporary occupancy is planned or the HN provides inadequate or no property. Early acquisition of facilities can be critical to the flow of forces. Use of automated planning tools can help forecast construction labor, materiel, and equipment requirements in support of the JFC's contingency basing plan.

(4) **Environmental Considerations.** Environmental considerations are broader than just protection of the environment and environmental stewardship. They also include continuously integrating the FHP, CMO, and other more operationally focused environmental considerations that affect US military forces and objectives. Military operations do not generally focus on environmental compliance and environmental protection. While complete protection of the environment will not always be possible due to its competition with other risks that the commander must assess, JFCs are to protect the environment in which US military forces operate to the greatest extent possible consistent with operational requirements. Specific planning guidance for environment issues should be IAW Chairman of the Joint Chiefs of Staff Manual (CJCSM) 3130.03, *Adaptive Planning and Execution (APEX) Formats and Guidance.* Commanders comply with the command guidance on environmental considerations specified in the plan or order and included in unit SOPs. Environmental considerations link directly to risk management and the safety and health of Service members. All significant risks must be clearly and accurately communicated to deploying DOD personnel and the chain of command. Environmental considerations, risk management, and health risk communication are enabling elements for the commander and are an essential part of military planning, training, and operations. While complete protection of the environment during military

Chapter III

operations may not always be possible, careful planning should address environmental considerations in joint operations, including legal aspects.

For additional guidance on environmental considerations, refer to DODI 4715.19, Use of Open-Air Burn Pits in Contingency Operations; *DODI 4715.22,* Environmental Management Policy for Contingency Locations; *JP 3-34,* Joint Engineer Operations; *and JP 4-02,* Joint Health Services.

(5) **Operational Energy.** The ability of the joint force to conduct operations depends on availability of sufficient energy, such as bulk fuel and electricity. Proper consideration of operational energy requirements improves the joint force's ability to maintain operational access. Furthermore, efficient management and use of operational energy may enable greater availability of combat forces for a variety of missions.

(6) **Health Services.** Health services promote, improve, preserve, or restore the behavioral or physical well-being of personnel. Health services include, but are not limited to, the management of health services resources, such as manpower, monies, and facilities; preventive and curative health measures; medical evacuation and patient movement of the sick, wounded, or injured; selection of the medically fit and disposition of the medically unfit; blood management; medical supply, equipment, and maintenance thereof; combat and operational stress control; and medical, dental, veterinary, laboratory, optometric, nutrition therapy, and medical intelligence services. Medical logistics, included within health services, includes patient movement, evacuation, and hospitalization. CCDRs are responsible for health services of forces assigned or attached to their command and should establish health services policies and programs.

(a) **Actions to obtain health threat information** begin prior to deployment and are continually updated as forces are deployed. Disease and injuries can quickly diminish combat effectiveness and have a greater impact on operations when the forces employed are small and dispersed.

(b) The **early introduction of preventive medicine personnel** or units into theater helps protect US forces from diseases and injuries. It also permits a thorough assessment of the health threat to and operational requirements of the mission. Preventive medicine support includes education and training on personal hygiene and field sanitation, personal protective measures, epidemiological investigations, pest management, and inspection of water sources and supplies. For maximum effectiveness, preventive medicine needs to be provided to as many personnel as possible within the OA. In addition to US forces, preventive medicine should include multinational forces, HN civilians, and dislocated civilians to the greatest feasible extent. JFCs and joint force surgeons shall identify legal constraints unique to the OE and intended recipient of services. Issues such as eligibility of beneficiaries, reimbursement for supplies and manpower, and provisions of legal agreements and other laws applicable to the theater, are reviewed.

(c) **Medical and rehabilitative care** provides essential care in the OA and rapid evacuation to definitive care facilities without sacrificing quality of care. It encompasses care provided from the point of illness or injury through rehabilitative care.

Joint Functions

For further guidance on health services, refer to JP 4-02, Joint Health Services. *For further guidance on procedures for deployment health activities, refer to DODI 6490.03,* Deployment Health.

(7) **HNS.** JFC's interact with the HN government to establish procedures to request support and negotiate support terms. Logistic planners should analyze the HN economic capacity to supplement the logistic support to US or multinational forces, and identify and limit adverse effects on the HN economy. Accordingly, early mission analysis should identify distribution requirements. This should be a collaborative analysis with HN government and private sector providers to build a systems analysis for designated focus areas when they are established. The systems analysis should evaluate airfields, seaports, rail and road networks, and energy infrastructure, particularly in underdeveloped countries where their status is questionable. Delaying this systems analysis can diminish the flow of strategic lift assets into the region. Additional support forces may be required to build or improve the supporting infrastructure to facilitate follow-on force closure as well as the delivery of humanitarian cargo.

(8) **OCS.** Logistic support requirements are often met through contracts with commercial entities inside and outside the OA. Most joint operations will require a level of contracted support. Certain contracted items or services could be essential to deploying, sustaining, and redeploying joint forces effectively. OCS is the process for obtaining supplies, services, and construction material from commercial sources in support of joint operations. OCS is a multi-faceted joint activity executed by the GCC and subordinate JFC's through boards, centers, working groups, and associated lead Service or joint theater support contract related activities. OCS includes the ability to plan, orchestrate, and synchronize the provision of contract support integration, contracting support, and contractor management.

Refer to JP 4-10, Operational Contract Support, *for further information on OCS.*

(9) **Disposal Operations.** Disposal is a consideration throughout planning, execution, and through redeployment. Inadequate understanding of disposal operations may cause violations of public and international law, confusion over roles and requirements, increased costs, inefficient operations, and negative health implications. Defense Logistics Agency (DLA) support to the CCDR's component commands includes the capability to receive and dispose of materiel in a theater. The DLA Disposition Services element in theater establishes theater-specific procedures for the reuse, demilitarization, or disposal of equipment and supplies, to include hazardous material and waste.

(10) **Legal Support.** Legal support is important across all joint functions. Many decisions and actions have potential legal implications. The JFC's staff judge advocate (SJA) provides the full spectrum of legal support during all joint operations through direct and reachback capability. A key member of the JFC's personal staff, the SJA provides legal advice on the laws, regulations, policies, treaties, and agreements that affect joint operations. Legal advisors actively participate in the planning process from mission analysis to execution, an essential function given the complexity of the OE. Legal representatives advise on fiscal activities, international law, and many other factors that

Chapter III

can affect operations, to include identifying legal issues that affect operational limitations. Further, the JFC should integrate HN legal personnel into the command legal staff as soon as practical to obtain guidance on unique HN legal practices and customs.

Refer to JP 1-04, Legal Support to Military Operations, for more detailed information and guidance on legal support.

(11) **Financial Management.** Financial management encompasses resource management and finance support. The joint force comptroller provides the elements of finance operations. The resource management normally consists of costing functions and leveraging fund sources. Finance operations provide funds to contract and limited pay support. The joint force comptroller's management of these elements provides the JFC with many capabilities, from contracting and banking support to cost capturing and fund control. Financial management support for contracting, subsistence, billeting, transportation, communications, labor, and a myriad of other supplies and services, particularly in austere environments, can enable mission accomplishment.

Refer to JP 1-06, Financial Management Support in Joint Operations, for more detailed information and guidance on financial management support.

CHAPTER IV
ORGANIZING FOR JOINT OPERATIONS

> *"Good will can make any organization work; conversely, the best organization chart in the world is unsound if the men who have to make it work don't believe in it."*
>
> **James Forrestal, Secretary of Defense 1947-1949**

1. Introduction

Organizing for joint operations involves many considerations. Most can be associated in three primary groups related to organizing the joint force, organizing the joint force HQ, and organizing OAs to help control operations. Understanding the OE helps the JFC understand factors that may affect decisions in each of these areas.

2. Understanding the Operational Environment

a. **General.** Factors that affect joint operations extend far beyond the boundaries of the JFC's assigned JOA. The JFC's **OE** is the composite of the conditions, circumstances, and influences that affect employment of capabilities and bear on the decisions of the commander. It encompasses physical areas of the air, land, maritime, and space domains; the information environment (which includes cyberspace); the EMS; and other factors. **Included within these are enemy, friendly, and neutral systems that are relevant to a specific joint operation.** The nature and interaction of these systems will affect how the commander plans, organizes for, and conducts joint operations.

b. **Physical Areas and Factors**

(1) **Physical Areas.** The fundamental physical area in the OE is the JFC's assigned OA. This term encompasses more descriptive terms for geographic areas in which joint forces conduct military operations. OAs include, but are not limited to, such descriptors as AOR, theater of war, theater of operations, JOA, AOA, joint special operations area (JSOA), and AO.

(2) **Physical Factors.** The JFC and staff must consider many factors associated with operations in the air, land, maritime, and space domains, and the information environment (which includes cyberspace). These factors include terrain (including urban settings), population, weather, topography, hydrology, EMS, and other environmental conditions in the OA; distances associated with the deployment to the OA and employment of joint capabilities; the location of bases, ports, and other supporting infrastructure; the physical results of combat operations; and both friendly and enemy forces and other capabilities. Combinations of these factors affect operations and sustainment.

c. **Information Environment.** The information environment is the aggregate of individuals, organizations, and systems that collect, process, disseminate, or act on information.

(1) The information environment is where humans and systems observe, orient, decide, and act upon information, and exists throughout the JFC's OE. The information environment consists of three interrelated dimensions—physical, informational, and cognitive—within which individuals, organizations, and systems continuously interact. Resources in this environment include the information itself and the materials and systems employed to process, store, display, disseminate, and protect information and produce information-related products.

(2) **Cyberspace** is a global domain within the information environment. It consists of the interdependent network of information technology infrastructures and resident data, including the Internet, telecommunications networks, computer systems, and embedded processors and controllers. Most aspects of joint operations rely in part on cyberspace, which reaches across geographic and geopolitical boundaries—much of it residing outside of US control—and is integrated with the operation of critical infrastructures, as well as the conduct of commerce, governance, and national security. Commanders must consider their critical dependencies on information and cyberspace, as well as factors such as degradations to confidentiality, availability, and integrity of information and information systems, when they plan and organize for operations.

(3) Commanders conduct CO to retain freedom of maneuver in cyberspace, accomplish the JFC's objectives, deny freedom of action to enemies and adversaries, and enable other operational activities. CO include DODIN operations to secure and operate DOD cyberspace. CO rely on links and nodes that reside in the physical domains, and perform functions in cyberspace and the physical domains. Similarly, activities in the physical domains can create effects in and through cyberspace by affecting the EMS or the physical infrastructure.

For more information on CO and the information environment, refer to JP 3-12, Cyberspace Operations, *and JP 3-13,* Information Operations.

d. **EMS.** The EMS is the range of all frequencies of electromagnetic radiation. Electromagnetic radiation consists of oscillating electric and magnetic fields characterized by frequency and wavelength. The EMS is usually subdivided into frequency bands based on certain physical characteristics and includes radio waves, microwaves, millimeter waves, infrared radiation, visible light, ultraviolet radiation, x-rays, and gamma rays. The rapid advances in EMS technologies over the last few decades have led to an exponential increase in commercial and military EMS-enabled/dependent capabilities. This proliferation, coupled with the US military's heavy reliance on the EMS and the low entry costs for adversaries, poses significant military challenges to the JFC. Integrated EMS operations are required in order to achieve success and EMS superiority—essential to all joint operations.

For more information on the EMS and EMS operations, see JP 3-13.1, Electronic Warfare, *and JP 6-01,* Joint Electromagnetic Spectrum Management Operations.

Organizing for Joint Operations

e. **A Systems Perspective**

(1) A **system** is a functionally, physically, or behaviorally related group of regularly interacting or interdependent elements forming a unified whole. One way to think of the OE is as a set of complex and constantly interacting political, military, economic, social, information, and infrastructure (PMESII) systems as depicted in Figure IV-1. The interaction of these systems can then be viewed as a network or networks based on the participants. The nature and interaction of these systems affect how the commander plans, organizes, and conducts joint operations. The JFC's intergovernmental partners and other civilian participants routinely focus on systems other than military, so the JFC and staff should understand these systems and how military operations affect them. Equally important is understanding how elements in other PMESII systems can help or hinder the

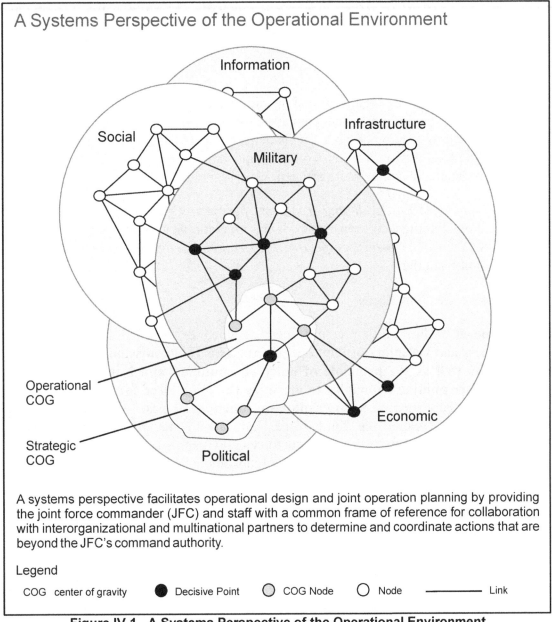

Figure IV-1. A Systems Perspective of the Operational Environment

IV-3

Chapter IV

JFC's mission. A commonly shared understanding among stakeholders in the operation can influence actions beyond the JFC's directive authority and promote a unified approach to achieve objectives.

(2) A systems understanding of the OE typically is built through cross-functional participation by other joint force staff elements and collaboration with various intelligence organizations, USG departments and agencies, and nongovernmental centers that possess expertise. The JFC should consider the best way to manage or support this cross-functional effort. The J-2 is the staff lead for this effort as part of JIPOE. A variety of factors, including planning time available, will affect the fidelity of a systems perspective.

(3) Understanding PMESII systems, their interaction with each other, and how system relationships will change over time will increase the JFC's knowledge of how actions within a system can affect other system components. Among other benefits, this perspective helps intelligence analysts identify potential sources of warning intelligence, and facilitates understanding the continuous and complex interaction of friendly, adversary, enemy, and neutral systems. A systems understanding also facilitates identification of operational design elements such as COGs, LOOs, and decisive points. For example, Figure IV-1 depicts notional operational and strategic COGs (there could be more). It shows each as a sub-system composed of related nodes and clarifies that the two COGs are related by a common node. This helps commanders and their staffs visualize and develop a broad approach to mission accomplishment early in the planning process, which makes detailed planning more efficient.

For further guidance on developing a systems perspective as part of JIPOE, refer to JP 2-01.3, Joint Intelligence Preparation of the Operational Environment. For further guidance on the use of a systems perspective in operational design and joint planning, refer to JP 5-0, Joint Planning.

3. **Organizing the Joint Force**

a. **General.** How JFCs organize their assigned or attached forces affects the responsiveness and versatility of joint operations. **The JFC's mission and operational approach, as well as the principle of unity of command and a mission command philosophy, are guiding principles to organize the joint force for operations.** Joint forces can be established on a **geographic or functional basis.** JFCs may centralize selected functions within the joint force, but should not reduce the versatility, responsiveness, and initiative of subordinate forces. JFCs should allow Service and special operations tactical and operational forces, organizations, and capabilities to function generally as they were designed. All Service components contribute distinct capabilities to joint operations that enable joint effectiveness. Joint interdependence is the purposeful reliance by one Service on another Service's capabilities to maximize the complementary and reinforcing effects of both. The degree of interdependence varies with specific circumstances. When JFCs organize their forces, they should also consider the degree of interoperability among Service components, with multinational forces and other potential participants. Complex or unclear command relationships are counterproductive to synergy among multinational forces. Simplicity and clarity of expression are essential. Similarly,

JFCs conducting domestic operations should consider overlapping responsibilities and interoperability issues among the Active Component and Reserve Components forces.

b. **Joint Force Options**

(1) **CCMDs.** A CCMD is a unified or specified command with a broad continuing mission under a single commander established and so designated by the President, through SecDef, and with the advice and assistance of the CJCS. **Unified commands** are established to conduct broad continuing missions requiring execution by significant forces of two or more Military Departments to achieve national objectives or other criteria found in JP 1, *Doctrine for the Armed Forces of the United States.* **Specified commands** are normally composed of forces from one Military Department, but may include units and staff representation from other Military Departments. The UCP establishes missions, responsibilities, and geographic AORs assigned to GCCs. A GCC is a CCDR assigned a geographic area that includes all associated land, water, and airspace. An FCC is a CCDR with transregional functional responsibilities. GCCs and FCCs have the authority to employ forces within their commands to carry out assigned missions, and they act as the supported commander for planning and executing of these missions. They may simultaneously be a supporting commander to other CCDRs for planning and executing the other CCDR's missions. In addition, US Special Operations Command and United States Transportation Command (USTRANSCOM) serve as joint force providers for SOF and mobility forces respectively.

(2) **Subordinate Unified Commands.** When authorized by SecDef through the CJCS, commanders of unified (not specified) commands may establish subordinate unified commands to conduct operations on a continuing basis IAW the criteria set forth for unified commands. A subordinate unified command may be established on a geographic area or functional basis. Commanders of subordinate unified commands have functions and responsibilities similar to those of the commanders of unified commands. They exercise OPCON of assigned commands and forces and normally of attached forces in the assigned operational or functional area.

(3) **Joint Task Forces (JTFs).** A JTF is a joint force constituted and designated by SecDef, a CCDR, a subordinate unified command commander, or an existing commander, joint task force (CJTF) to accomplish missions with specific, limited objectives, and which do not require centralized control of logistics. However, there may be situations where a CJTF may require directive authority for common support capabilities delegated by the CCDR. JTFs may be established on a geographical area or functional basis. However, JTFs can also be established based on a security challenge that focuses on specific threats that cross AOR boundaries or multiple noncontiguous geographic areas. The proper authority dissolves a JTF when the JTF achieves the objectives for which it was created or is no longer required.

(4) There are several ways to form a JTF HQ. Normally, a CCMD may employ a Service component HQ or one of the Service component's existing subordinate HQ (e.g., Army corps, numbered air force, numbered fleet and Marine expeditionary force) as the core of a JTF HQ and then augment that core with personnel and capabilities from the

Chapter IV

> **COMMON OPERATING PRECEPT**
>
> **Maintain operational and organizational flexibility.**

Services comprising the JTF. Also, the theater special operations command (TSOC) or a subordinate SOF HQ with the C2 capability can form the foundation for a JTF HQ. CCDRs verify the readiness of assigned Service HQ staffs to establish, organize, and operate as a JTF-capable HQ. JTF HQ basing depends on the JTF mission, OE, and available capabilities and support. JTF HQ can be land- or sea-based, with transitions between both basing options. JTFs are normally assigned a JOA. JTFs must be able to integrate effectively with USG departments and agencies, multinational partners, and indigenous and regional stakeholders. When direct participation by USG departments and agencies other than DOD is significant, the TF establishing authority may designate it as a joint interagency TF. This typically occurs when the other interagency partners have primacy and legal authority and the JFC provides supporting capabilities, such as humanitarian assistance.

(5) Forming and training the joint force HQ and task organizing the joint force can be challenging, particularly in crisis action situations. Joint forces must quickly adjust both operations and organization in response to planned operational transitions or unexpected situational transitions. For example, achieving combat objectives in the dominate phase of an operation much earlier than anticipated could signal to the JFC to shift emphasis and organization quickly to stability actions commonly associated with the stabilize and enable civil authority phases. Similarly, the JFC's mission will affect the echelon at which joint capabilities are best employed. Advances in areas ranging from communications and information sharing to munitions effectiveness make it possible to synchronize lower echelons of command in some situations without the risks and inefficiencies associated with fragmenting the assets themselves. JFCs should exploit such opportunities.

> **COMMON OPERATING PRECEPT**
>
> **Drive synergy to the lowest echelon at which it can be managed effectively.**

For further guidance on the formation and employment of a JTF HQ to command and control a joint operation, refer to JP 3-33, Joint Task Force Headquarters.

c. **Component Options.** CCDRs and subordinate unified commanders conduct either single-Service or joint operations to accomplish a mission. All JFCs may conduct operations through their Service component commanders, lower-echelon Service force commanders, and functional component commanders. Further, **functional and Service components of the joint force conduct supported, subordinate, and supporting operations, not independent campaigns.**

(1) **Service Components.** Regardless of the organization and command arrangements within joint commands, Service component commanders retain responsibility for certain Service-specific functions and other matters affecting their forces, including internal administration, personnel support, training, sustainment, and Service intelligence operations. Conducting joint operations through Service components has certain advantages, which include clear and uncomplicated command lines. This arrangement is appropriate when stability, continuity, economy, ease of long-range planning, and scope of operations dictate organizational integrity of Service components. While sustainment remains a Service responsibility, there are exceptions such as arrangements described in Service support agreements, CCDR-directed common-user logistics lead Service, or DOD agency responsibilities.

(2) **Functional Components.** The JFC can establish functional component commands to conduct operations when forces from two or more Services must operate in the same physical domain or accomplish a distinct aspect of the assigned mission. These conditions apply when the scope of operations requires that the similar capabilities and functions of forces from more than one Service be directed toward closely related objectives and unity of command is a primary consideration. For example, functionally oriented components are useful when the scope of operations is large and the JFC's attention must be divided between major operations or phases of operations that are functionally dominated. Functional component commands are subordinate components of a joint force. Except for the joint force special operations component and joint special operations task force, functional components do not constitute a joint force with a JFC's authorities and responsibilities, even when composed of forces from two or more Military Departments.

(a) JFCs may conduct operations through functional components or employ them primarily to coordinate selected functions. The JFC will normally designate the Service component commander who has the preponderance of forces and the ability to exercise C2 over them as the functional component commander. However, the JFC will always consider the mission, nature and duration of the operation, force capabilities, and C2 capabilities when selecting a commander. The establishment of a functional component commander must not affect the command relationship between Service component commanders and the JFC.

(b) The functional component commander's staff composition should reflect the command's composition so the staff has the required expertise to help the commander effectively employ the component's forces. Functional component staffs require advanced planning, appropriate training, and frequent exercises for efficient operations. Liaison elements from and to other components facilitate coordination and support. Staff billets and individuals to fill them should be identified and used when the commander forms the functional component staff for exercises and actual operations. The number of staff personnel should be appropriate for the mission and nature of the operation. The staff structure should be flexible enough to add or delete personnel and capabilities in changing conditions without losing effectiveness.

Chapter IV

(c) The JFC designates forces and/or military capabilities that will be made available for tasking by the functional component commander and the appropriate command relationship(s). JFCs may also establish a support relationship between components to facilitate operations. Regardless, the establishing JFC defines the authority and responsibilities of functional component commanders based on the CONOPS, and the JFC may alter their authority and responsibilities during the course of an operation.

(3) **Combination.** Joint forces often are organized with a combination of Service and functional components. For example, joint forces organized with Service components normally have SOF organized under a JFSOCC, while the conventional air forces will normally have a JFACC designated, whose authorities and responsibilities are defined by the establishing JFC based on the JFC's CONOPS.

d. **SOF Employment Options**

(1) Used independently with conventional force support (since USC limits SOF combat support and combat service support) or integrated with conventional forces, SOF provide strategic options for national leaders and the GCCs through a global network that fully integrates military, interagency, and international partners. SOF are most effective when special operations are fully integrated into the overall plan, and the execution of special operations is through proper SOF C2 elements employed intact.

(2) Commander, United States Special Operations Command (CDRUSSOCOM) synchronizes the planning of special operations and provides SOF to support persistent, networked, and distributed GCC operations to protect and advance national interests.

(3) CDRUSSOCOM exercises COCOM of all SOF. GCCs exercise OPCON of their supporting TSOCs and most often exercise OPCON of SOF deployed in their AORs. The establishing directive will define command relationships between the special operations commands and JTF/TF commanders. A TSOC commander can be the JTF commander.

For more information on special operations, refer to JP 3-05, Special Operations.

e. **Joint HQ Augmentation Options.** There are various options available to augment a joint HQ that is forming for joint operations.

See JP 3-33, Joint Task Force Headquarters, for more information.

4. **Organizing the Joint Force Headquarters**

a. Joint force HQ include those for unified, subordinate unified, and specified commands and JTFs. While each HQ organizes to accommodate the nature of the JFC's OA, mission, tasks, and preferences, all generally follow a traditional functional staff alignment (i.e., personnel, intelligence, operations, logistics, plans, and communications) depicted in Figure IV-2. The primary staff functional areas are also generally consistent with those at Service component HQ, which facilitates higher, lower, and lateral cross-command staff coordination and collaboration. Some HQ may combine functions under a

Organizing for Joint Operations

staff principal, while other HQ may add staff principals. Based on the mission and other factors, some joint HQ may have additional primary staff organizations focused on areas such as engineering; force structure, resource, and assessment; and CMO.

b. Figure IV-2 also shows boards, centers, working groups, and other semi-permanent and temporary organizations. These facilitate cross-functional coordination, synchronization, planning, and information sharing between principal staff directorates. Although these organizations are cross-functional in their membership, they typically fall under the oversight of a principal staff directorate or senior staff member. For example, the joint operations center aligns under the J-3, the joint intelligence support element under the J-2, and the joint media operations center under PA.

c. HQ also have personal and special staff sections or elements, which perform specialized duties as prescribed by the JFC and handle special matters over which the JFC wishes to exercise personal control. Examples include the SJA, provost marshal, and inspector general.

For detailed guidance on the organization of a joint force HQ, refer to JP 3-33, Joint Task Force Headquarters.

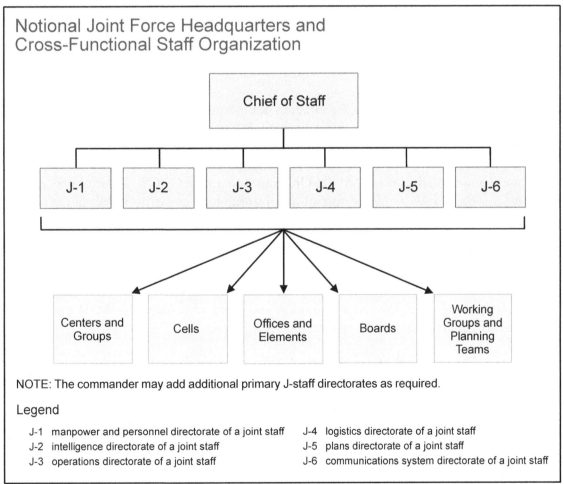

Figure IV-2. Notional Joint Force Headquarters and Cross-Functional Staff Organization

Chapter IV

> **OPERATIONAL AREAS FOR OPERATION RESTORE HOPE**
>
> During Operation RESTORE HOPE in Somalia, the joint forces rear area was centered around the separate sites of the embassy compound, port, and airfield in the city of Mogadishu, while its operational area was widely scattered around the towns and villages of the interior. The area of interest included the rest of the country and particularly those population and relief centers not under the joint force commander's supervision.
>
> **Various Sources**

5. Organizing Operational Areas

a. **General.** Except for AORs, which are assigned in the UCP, GCCs and other JFCs designate smaller operational areas (e.g., JOA and AO) on a temporary basis. OAs have physical dimensions comprised of some combination of air, land, maritime, and space domains. While domains are useful constructs for visualizing and characterizing the physical environment in which operations are conducted (the OA), the use of the term "domain" is not meant to imply or mandate exclusivity, primacy, or C2 of any domain. Specific authorities and responsibilities within an operational area are as specified by the appropriate JFC. JFCs define these areas with geographical boundaries, which help commanders and staffs coordinate, integrate, and deconflict joint operations among joint force components and supporting commands. The size of these OAs and the types of forces employed within them depend on the scope and nature of the mission and the projected duration of operations.

b. **CCMD-Level Areas.** GCCs conduct operations in their assigned AORs. When warranted, the President, SecDef, or GCCs may designate a theater of war and/or theater of operations for each operation (see Figure IV-3). GCCs can elect to control operations directly in these OAs, or may establish subordinate joint forces for that purpose, while remaining focused on the broader AOR. Operations that span GCC boundaries may expose gaps in C2. DOD uses a mix of formal and informal processes to synchronize actions between AORs.

(1) **AOR.** An AOR is an area established by the UCP that defines geographic responsibilities for a GCC. A GCC has authority to plan for operations within the AOR and conduct those operations approved by the President or SecDef. CCDRs may operate forces wherever required to accomplish approved missions. **All cross-AOR operations must be coordinated among the affected GCCs.**

(2) **Theater of War.** A theater of war is a geographical area established by the President, SecDef, or GCC for the conduct of major operations and campaigns involving combat. A theater of war is established primarily when there is a formal declaration of war or it is necessary to encompass more than one theater of operations (or a JOA and a separate theater of operations) within a single boundary for the purposes of C2, sustainment,

Organizing for Joint Operations

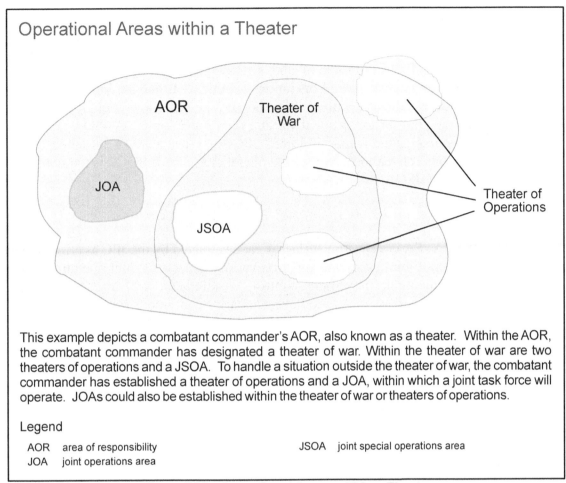

Figure IV-3. Operational Areas within a Theater

protection, or mutual support. A theater of war does not normally encompass a GCC's entire AOR, but may cross the boundaries of two or more AORs.

(3) **Theater of Operations.** A theater of operations is an OA defined by the GCC for the conduct or support of specific military operations. A theater of operations is established primarily when the scope of the operation in time, space, purpose, and/or employed forces exceeds what a JOA can normally accommodate. More than one joint force HQ can exist in a theater of operations. A GCC may establish one or more theaters of operations. Different theaters will normally be focused on different missions. A theater of operations typically is smaller than a theater of war, but is large enough to allow for operations in depth and over extended periods of time. Theaters of operations are normally associated with major operations and campaigns and may cross the boundary of two AORs.

c. For operations somewhat limited in scope and duration, or for specialized activities, the commander can establish the following OAs.

(1) **JOA.** A JOA is an area of land, sea, and airspace, defined by a GCC or subordinate unified commander, in which a JFC (normally a CJTF) conducts military operations to accomplish a specific mission. JOAs are particularly useful when operations

Chapter IV

are limited in scope and geographic area or when operations are to be conducted on the boundaries between theaters.

(2) **JSOA.** A JSOA is an area of land, sea, and airspace assigned by a JFC to the commander of SOF to conduct special operations activities. It may be limited in size to accommodate a discreet direct action mission or may be extensive enough to allow a continuing broad range of unconventional warfare (UW) operations. A JSOA is defined by a JFC who has geographic responsibilities. JFCs may use a JSOA to delineate and facilitate simultaneous conventional and special operations. The JFSOCC is the supported commander within the JSOA.

For additional guidance on JSOAs, refer to JP 3-05, Special Operations.

(3) **JSA.** A JSA is a specific surface area, designated by the JFC as critical, that facilitates protection of joint bases and supports various aspects of joint operations such as LOCs, force projection, movement control, sustainment, C2, airbases/airfields, seaports, and other activities. JSAs are not necessarily contiguous with areas actively engaged in combat (see Figure IV-4). JSAs may include intermediate support bases and

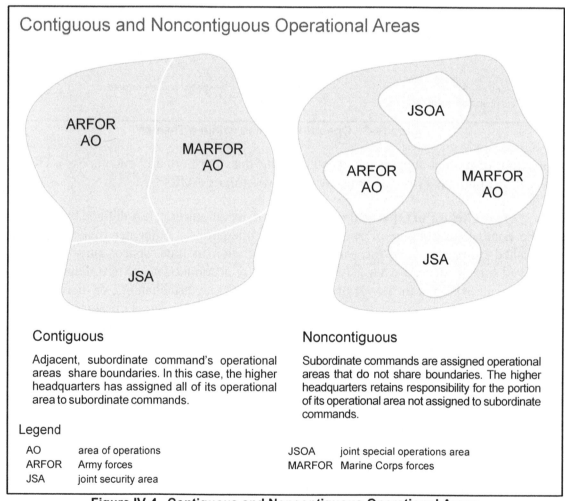

Figure IV-4. Contiguous and Noncontiguous Operational Areas

other support facilities intermixed with combat elements. JSAs may be used in both linear and nonlinear situations.

For additional guidance on JSAs, refer to JP 3-10, Joint Security Operations in Theater.

(4) **AOA.** The AOA is a geographic area within which is located the objective(s) to be secured by the amphibious force. This area must be of sufficient size to ensure accomplishment of the amphibious force's mission and must provide sufficient area for conducting necessary sea, air, and land operations.

For additional guidance on AOAs, refer to JP 3-02, Amphibious Operations.

(5) **AO.** JFCs may define AOs for land and maritime forces. AOs do not typically encompass the entire OA of the JFC, but should be large enough for component commanders to accomplish their missions (to include a designated amount of airspace) and protect their forces. Component commanders with AOs typically designate subordinate AOs within which their subordinate forces operate. These commanders employ the full range of joint and Service control measures and graphics as coordinated with other component commanders and their representatives to delineate responsibilities, deconflict operations, and achieve unity of effort.

d. **Contiguous and Noncontiguous OAs**

(1) OAs may be contiguous or noncontiguous (Figure IV-4). When they are contiguous, a boundary separates them. When OAs are noncontiguous, subordinate commands do not share a boundary. The higher HQ retains responsibility for the unassigned portion of its OA.

(2) In some operations, a Service or functional component (typically the ground component) could have such a large OA that the component's subordinate units operate in a noncontiguous manner, widely distributed and beyond mutually supporting range of each other. In these cases, the JFC should consider options whereby joint capabilities can be pushed to lower levels and placed under control of units that can use them effectively.

e. **Considerations When Assuming Responsibility for an OA.** The establishing commander should activate an assigned OA at a specified date and time based on mission and situation considerations addressed during COA analysis and wargaming. Among others, common considerations include C2, the information environment, intelligence requirements, communications support, protection, security, LOCs, terrain management, movement control, airspace control, surveillance, reconnaissance, air and missile defense, PR, targeting and fires, interorganizational coordination, and environmental issues.

Refer to JP 3-33, Joint Task Force Headquarters, *for specific guidance on assuming responsibility for an OA.*

Intentionally Blank

CHAPTER V
JOINT OPERATIONS ACROSS THE CONFLICT CONTINUUM

> *"I am confident that our Nation has the most professional and capable military in the world. Our Joint Forces have proven effective and resilient throughout years of combat, kept the homeland safe, and advanced our national interests across the globe."*
>
> **General Joseph F. Dunford, Jr., 19th Chairman of the Joint Chiefs of Staff,** *Joint Force Quarterly 80*

1. Introduction

a. Threats to US and allied interests throughout the world can sometimes only be countered by US forces able to respond to a wide variety of challenges along a **conflict continuum** that spans from peace to war. Our national interests and the nature of crises that can occur along this continuum require our nation's armed forces to be proficient in a wide variety of activities, tasks, missions, and operations that vary in purpose, scale, risk, and combat intensity.

b. As the joint operations keystone publication, JP 3-0 provides a broad framework within which to consider how to conduct joint operations. The framework comprises a combination of fundamental constructs, such as **operational art, the range of military operations,** and the **interconnected OE. Together, they provide broad context to consider the use of military capabilities in various circumstances across the conflict continuum.** This chapter discusses some of these constructs to clarify the relationship of various military operations and activities to each other, and in the context of the OE.

2. Military Operations and Related Missions, Tasks, and Actions

a. In general, a military operation is a set of actions intended to accomplish a task or mission. Although the US military is organized, trained, and equipped for sustained, large-scale combat anywhere in the world, the capabilities to conduct these operations also enable a wide variety of other operations and activities. In particular, opportunities exist prior to large-scale combat to shape the OE in order to prevent, or at least mitigate, the effects of war. Characterizing the employment of military capabilities (people, organizations, and equipment) as one or another type of military operation has several benefits. For example, publications can be developed that describe the nature, tasks, and tactics associated with specific types of diverse operations, such as NEO and COIN. These publications provide the basis for related joint training and joint professional military education that help joint forces conduct military operations as effectively and efficiently as possible even in difficult and dangerous circumstances. Characterizations also help military and civilian leaders explain US military involvement in various situations to the US and international public and news media. Likewise, such characterizations, supplemented by operational experience, can clarify the need for specific capabilities that enhance certain operations. For example, facial recognition software associated with

Chapter V

biometric capabilities helps military and law enforcement personnel identify terrorists and piece together their human networks as part of combating terrorism.

Refer to JP 3-07.2, Antiterrorism, *for additional guidance on antiterrorism, and JP 3-26,* Counterterrorism, *for additional guidance on counterterrorism.*

b. Military operations are often categorized by their focus, as shown in Figure V-1. In some cases, the title covers a variety of missions, tasks, and activities. Many activities accomplished by single Services, such as tasks associated with security cooperation, do not constitute a joint operation. Nonetheless, most of these occur under a joint "umbrella," because they contribute to achievement of CCDRs' TCP objectives. Following are brief summaries of examples of military operations and activities.

(1) **Stability Activities.** Stability activities is an overarching term encompassing various military missions, tasks, and activities conducted outside the US in coordination with other instruments of national power to maintain or reestablish a safe and secure environment and to provide essential governmental services, emergency infrastructure reconstruction, and humanitarian relief. See JP 3-07, *Stability,* for more information.

(2) **DSCA.** DSCA is provided by US federal military forces, DOD agencies, DOD civilians, DOD contract personnel, DOD component assets, and National Guard forces, when SecDef, in coordination with the governors of the affected states, elects and requests to use those forces in Title 32, USC, status, in response to requests for assistance from civil authorities for domestic emergencies, law enforcement support, and other domestic activities, or from qualifying entities for special events. See JP 3-28, *Defense Support of Civil Authorities,* for more information.

(3) **Foreign Humanitarian Assistance (FHA).** FHA is DOD activities, normally in support of the United States Agency for International Development (USAID) or DOS, conducted outside the US and its territories to relieve or reduce human suffering, disease, hunger, or privation. See JP 3-29, *Foreign Humanitarian Assistance.*

Examples of Military Operations and Activities

- Stability activities
- Defense support of civil authorities
- Foreign humanitarian assistance
- Recovery
- Noncombatant evacuation
- Peace operations
- Countering weapons of mass destruction
- Chemical, biological, radiological, and nuclear response
- Foreign internal defense
- Counterdrug operations
- Combating terrorism
- Counterinsurgency
- Homeland defense
- Mass atrocity response
- Security cooperation
- Military engagement

Figure V-1. Examples of Military Operations and Activities

Joint Operations Across the Conflict Continuum

(4) **Recovery.** Recovery is operations to search for, locate, identify, recover, and return isolated personnel, human remains, sensitive equipment, or items critical to national security. See JP 3-50, *Personnel Recovery.*

(5) **NEO.** NEO is an operation to evacuate noncombatants and civilians from foreign countries to safe havens or to the US when their lives are endangered by war, civil unrest, or natural disaster. See JP 3-68, *Noncombatant Evacuation Operations.*

(6) **Peace Operations (PO).** PO are operations to contain conflict, redress the peace, and shape the environment to support reconciliation and rebuilding and facilitate the transition to legitimate governance. PO include peacekeeping operations (PKO), peace enforcement operations (PEO), peacemaking (PM), peace building (PB), and conflict prevention efforts. See JP 3-07.3, *Peace Operations.*

(7) **Countering Weapons of Mass Destruction (CWMD).** CWMD encompasses efforts against actors of concern to curtail the conceptualization, development, possession, proliferation, use, and effects of WMD, related expertise, materials, technologies, and means of delivery. See JP 3-40, *Countering Weapons of Mass Destruction.*

(8) **CBRN Response.** CBRN response is DOD support to USG actions that plan for, prepare for, respond to, and recover from the effects of domestic and foreign CBRN incidents. See JP 3-41, *Chemical, Biological, Radiological, and Nuclear Response.*

(9) **Foreign Internal Defense (FID).** FID is participation by civilian and military agencies of a government in any of the action programs taken by another government or other designated organization to free and protect its society from subversion, lawlessness, insurgency, terrorism, and other threats to its security. FID is an example of USG foreign assistance. See JP 3-22, *Foreign Internal Defense.*

(10) **Counterdrug (CD) Operations.** CD operations provide DOD support to LEAs to detect, monitor, and counter the production, trafficking, and use of illegal drugs. See JP 3-07.4, *Counterdrug Operations.*

(11) **Combating Terrorism.** Combatting terrorism is actions, including antiterrorism (defensive measures taken to reduce vulnerability to terrorist acts) and counterterrorism (CT) (actions taken directly against terrorist networks) to oppose terrorism. See JP 3-07.2, *Antiterrorism,* and JP 3-26, *Counterterrorism.*

(12) **COIN.** COIN is an operation that encompasses comprehensive civilian and military efforts taken to defeat an insurgency and to address any core grievances. See JP 3-24, *Counterinsurgency Operations.*

(13) **HD.** HD is the protection of US sovereignty, territory, domestic population, and critical defense infrastructure against external threats and aggression or other threats as directed by the President. See JP 3-27, *Homeland Defense.*

Chapter V

(14) **Mass Atrocity Response.** Mass atrocity response is military activities conducted to prevent or halt mass atrocities. See JP 3-07.3, *Peace Operations.*

3. **The Range of Military Operations**

a. The **range of military operations** is a fundamental construct that helps relate military activities and operations in scope and purpose. The potential range of military activities and operations extends from military engagement, security cooperation, and deterrence in times of relative peace up through large-scale combat operations. The range encompasses three primary categories: **military engagement, security cooperation, and deterrence; crisis response and limited contingency operations; and large-scale combat operations.** Figure V-2 depicts these categories against a backdrop of the conflict continuum. All operations across this range share a common fundamental purpose—to achieve or contribute to national objectives.

b. Military engagement, security cooperation, and deterrence activities develop local and regional situational awareness, build networks and relationships with partners, shape the OE, keep day-to-day tensions between nations or groups below the threshold of armed conflict, and maintain US global influence. Many missions associated with crisis response and limited contingencies, such as DSCA and FHA, may not require combat. But others, such as Operation RESTORE HOPE in Somalia, can be dangerous and may require combat operations to protect US forces. Large-scale combat often occurs in the form of major operations and campaigns that achieve national objectives or contribute to a larger, long-term effort (e.g., OEF).

c. The complex nature of the strategic environment may require US forces to conduct different types of joint operations and activities simultaneously across the conflict continuum. Although this publication discusses specific types of operations and activities under the various categories in the range of military operations, each type is not doctrinally

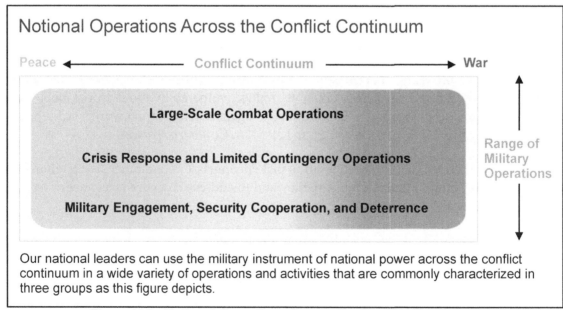

Figure V-2. Notional Operations Across the Conflict Continuum

fixed and could shift within that range. For instance, security cooperation activities may be satisfying internal security requirements of a PN, but the emergence of a violent internal security threat that overwhelms PN security forces could require the USG to commit to FID or COIN operations in that PN, while security cooperation activities continue.

(1) **Military Engagement, Security Cooperation, and Deterrence.** These ongoing activities establish, shape, maintain, and refine relations with other nations and include military engagement activities with domestic civil authorities (e.g., state governors or local law enforcement). The general strategic objective is to protect or further US interests at home and abroad by enabling support from PNs, enhancing their capacity or capability for security and stability, and maintaining or establishing operational access. These activities seek to build networks and relationships to develop situational and cultural awareness that allows the US to develop more informed options to address emerging situations and opportunities. These occur continuously in many parts of the GCC's AOR even when combat operations may be occurring in other parts. Some activities, such as those related to deterrence, can span the entire conflict continuum, and may be integral tasks associated with flexible deterrent options (FDOs), flexible response options (FROs), or large-scale combat operations. See Chapter VI, "Military Engagement, Security Cooperation, and Deterrence."

(2) **Crisis Response and Limited Contingency Operations.** These operations can range from an independent, small-scale, noncombat operation, such as support of civil authorities, up to a supporting component of extended major noncombat and/or combat operations. The associated general strategic and operational-level objectives are to **protect** US interests and/or **prevent** further escalation. See Chapter VII, "Crisis Response and Limited Contingency Operations."

(3) **Large-Scale Combat Operations.** The nature and scope of some missions may require joint forces to conduct large-scale combat operations to achieve national strategic objectives or protect national interests. Such combat typically occurs within the framework of a major operation or campaign. A major operation is series of tactical actions (battles, engagements, strikes) conducted by combat forces of a single or several Services, coordinated in time and place, to achieve strategic or operational objectives in an OA. The term can also refer to a noncombat operations of significant size and scope. A campaign is a series of related major operations aimed at achieving strategic and operational objectives within a given time and space. Usually associated with large-scale combat, a campaign also can comprise predominately limited combat and noncombat operations of extended duration to achieve theater and national strategic objectives. See Chapter VIII, "Major Operations and Campaigns."

4. **The Theater Campaign**

a. Military operations, actions, and activities in a GCC's AOR, from security cooperation through large-scale combat, are conducted in the context of the GCC's ongoing theater campaign.

Chapter V

b. The CCDR's theater campaign is the overarching framework that ensures all activities and operations within the theater are synchronized to achieve theater and national strategic objectives. A TCP operationalizes the GCC's strategy and approach to achieve these objectives within two to five years by organizing and aligning available resources. TCPs also support the campaign objectives of other CCDRs responsible for synchronizing collaborative DOD planning. As Figure V-3 shows, TCPs encompass all ongoing and planned operations across the range of military operations, continuously adjusted in response to changes in the OE. The TCP's long-term and persistent and preventative activities are intended to identify and deter, counter, or otherwise mitigate an adversary's actions before escalation to combat. Many of these activities are conducted with DOD in support of the diplomatic, economic, and informational efforts of USG partners and PNs. The CCDR adjusts these activities as required for the occasional execution of a contingency plan or response to a crisis.

For more information on theater campaign plans, see JP 5-0, Joint Planning.

c. The TCP also provides context for ongoing crisis response and contingency operations to facilitate execution of contingency plans as branch plans to the TCP. These are plans to respond to potential crises such as natural or man-made disasters and military

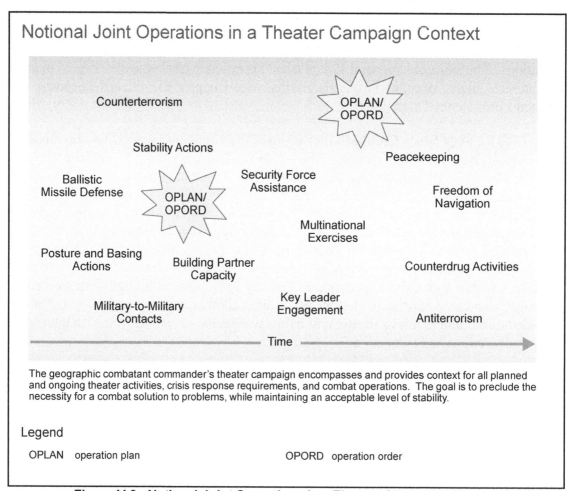

Figure V-3. Notional Joint Operations in a Theater Campaign Context

Joint Operations Across the Conflict Continuum

aggression by foreign powers. Also linked to the GCC's TCP and subordinate campaign plans are designated DOD global campaign plans that address integrated execution of global security priorities.

d. GCCs prepare TCPs to achieve military objectives and GEF-directed prioritized campaign and DOD objectives for their AORs. GCCs integrate planning in their TCPs to support designated missions assigned to the CCDRs responsible for collaborative DOD planning. Similar to GCCs, FCCs prepare global campaigns to achieve global security priorities as required in the GEF. FCCs also prepare functional campaign plans to achieve military objectives and GEF-directed objectives for their missions. FCCs synchronize planning for designated missions across CCMDs, Services, and DOD agencies.

e. **A GCC can simultaneously conduct multiple joint operations with different objectives within their AOR.** The GCC might initiate one or more OPLANs while security cooperation activities are ongoing in the same or another part of the theater. Further, a crisis response or contingency operation could occur separately or as part of a campaign or major operation (e.g., the NEO in Somalia during Operation DESERT SHIELD in 1991). In the extreme, a major operation or a subordinate campaign may occur within a theater concurrently with a separate or related campaign. **CCDRs should synchronize and integrate the activities of assigned, attached, and allocated forces with subordinate and supporting JFCs** so they complement rather than compete in achieving national or theater-strategic objectives. Due to the transregional nature of various enemies or adversaries (e.g., insurgents, terrorists, drug cartels, pirates), coordination and synchronization requirements may also extend to adjacent GCCs. CCDRs and subordinate JFCs must work with DOS regional and functional bureaus, individual country chiefs of mission, and other USG departments and agencies to better integrate military operations in unified action with the diplomatic, economic, and informational instruments of national power.

f. **Some military operations may be conducted for one purpose.** For example, FHA is focused on a humanitarian purpose (e.g., Operation TOMODACHI, an assistance operation to support Japan in disaster relief following the 2011 Tohoku earthquake and tsunami). A strike may be conducted for the specific purpose of compelling or deterring an action (e.g., Operation EL DORADO CANYON, the 1986 operation to coerce Libya to conform with international laws against terrorism). Often, however, military operations will have multiple purposes (based on strategic and operational-level objectives) and will be influenced by a fluid and changing situation. Branch and sequel events may produce additional tasks for the force, challenging the command with multiple missions (e.g., Operations PROVIDE COMFORT in Iraq and RESTORE HOPE in Somalia were PEO that evolved from FHA efforts). Joint forces must strive to meet such challenges with clearly defined objectives addressing diverse purposes.

5. A Joint Operation Model

a. Most individual joint operations share certain activities or actions in common. These include forming a joint HQ, deploying and redeploying capabilities (forces, materiel, etc.), and interacting with other interorganizational participants. Some activities can also

Chapter V

characterize specific types of operations such as large-scale combat, FHA, and COIN. For example, Figure V-4 shows six general groups of military activities that may typically occur in preparation for and during a single large-scale joint combat operation.

b. The nature of operations and activities during a typical joint combat operation will change from its beginning (when the CJCS issues the execute order) to the operation's end (when the joint force disbands and components return to a pre-operation status). **Shaping** activities usually precede the operation and may continue during and after the operation. The purpose of **shaping** activities is to help set the conditions for successful execution of the operation. Figure V-4 shows that from deter through enable civil authority, the operations and activities in these groups vary in magnitude—time, intensity, forces, etc.— as the operation progresses (the relative magnitudes in the figure are notional). At various points in time, each specific group might characterize the main effort of the joint force.

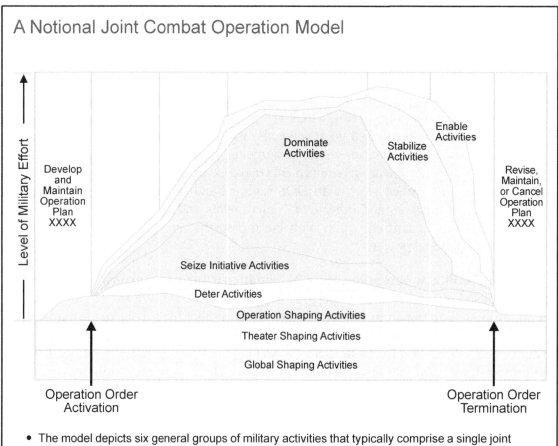

Figure V-4. A Notional Joint Combat Operation Model

Joint Operations Across the Conflict Continuum

For example, **dominate** activities would characterize the main effort after the joint force **seizes the initiative** until the enemy no longer is able to effectively resist. Even so, activities in the other groups would usually occur concurrently at some level of effort. The following paragraphs provide more information on the nature of these activities.

(1) **Shape**

(a) In general, shaping activities help set conditions for successful theater operations. Shaping activities include long-term persistent and preventive military engagement, security cooperation, and deterrence actions to assure friends, build partner capacity and capability, and promote regional stability. They help identify, deter, counter, and/or mitigate competitor and adversary actions that challenge country and regional stability. A GCC's TCP provides these and other activities tasked by SecDef/CJCS strategic guidance in pursuit of national objectives. Likewise, CCDRs may direct more focused geographic and functional shaping activities at the potential execution of specific contingency plans for various types of operations. In the best case, shaping activities may avert or diminish conflict. At the least, shaping provides a deeper, and common, understanding of the OE. Preparatory intelligence activities inform operation assessment, planning, and execution to improve the JFC's understanding of the OE.

(b) Shaping activities are largely conducted through other interorganizational participants (e.g., USG departments and agencies, PNs), with DOD in a supporting role. Where US and PN interests converge, cooperation is possible. Some partners are quite capable already; others may benefit from US assistance. When a nation shares our interests and has the capacity to absorb US training, regional security can be increased. Military engagement and security cooperation activities are executed continuously to enhance international legitimacy and gain multinational cooperation. These activities should improve perceptions and influence adversaries' and allies' behavior, develop allied and friendly military capabilities for self-defense and multinational operations, improve information exchange and intelligence sharing, provide US forces with peacetime and contingency access, and positively affect conditions that could lead to a crisis. These activities prepare the OE in advance to facilitate access, should contingency operations be required. The joint community, in concert with multinational and interagency partners, must maintain and exercise strong regional partnerships as essential shaping activities in peacetime to ensure operational access during plan execution. For example, obtaining and maintaining rights of navigation and overflight help ensure global reach and rapid projection of military power.

(2) **Deter.** Successful deterrence prevents an adversary's undesirable actions, because the adversary perceives an unacceptable risk or cost of acting. Deterrent actions are generally weighted toward protection and security activities that are characterized by preparatory actions to protect friendly forces, assets, and partners, and indicate the intent to execute subsequent phases of the planned operation. A number of FDOs, FROs, and force enhancements could be implemented during this phase. The nature of these options varies according to the nature of the adversary (e.g., traditional or irregular, state or non-state), the adversary's actions, US national objectives, and other factors. Once a crisis is defined, these actions may include mobilization, tailoring of forces, and other

Chapter V

predeployment activities; initial deployment into a theater; employment of intelligence collection assets; and development of mission-tailored C2, intelligence, force protection, and logistic requirements to support the JFC's CONOPS. CCDRs continue to conduct military engagement with multinational partners to maintain access to areas, thereby providing the basis for further crisis response. Many deterrent actions build on security cooperation activities. They can also be part of stand-alone operations.

(3) **Seize Initiative.** JFCs seek to seize the initiative in all situations through decisive use of joint force capabilities. In combat, this involves both defensive and offensive operations at the earliest possible time, forcing the enemy to culminate offensively and setting the conditions for decisive operations. Rapid application of joint combat power may be required to delay, impede, or halt the enemy's initial aggression and to deny the enemy its initial objectives. Operations to gain access to theater infrastructure and expand friendly freedom of action continue during this phase, while the JFC seeks to degrade enemy capabilities with the intent of resolving the crisis at the earliest opportunity.

(4) **Dominate.** These actions focus on breaking the enemy's will to resist or, in noncombat situations, to control the OE. Successful domination depends on overmatching enemy capabilities at critical times and places. Joint force options include attacking weaknesses at the leading edge of the enemy's defensive perimeter to roll enemy forces back, and striking in depth to threaten the integrity of the enemy's A2/AD, offensive weapons and force projection capabilities, and defensive systems. Operations can range from large-scale combat to various stability actions depending on the nature of the enemy. Dominating activities may establish the conditions to achieve strategic objectives early or may set the conditions for transition to a subsequent phase of the operation.

(5) **Stabilize.** These actions and activities are typically characterized by a shift in focus from sustained combat operations to stability activities. These operations help reestablish a safe and secure environment and provide essential government services, emergency infrastructure reconstruction, and humanitarian relief. The intent is to help restore local political, economic, and infrastructure stability. Civilian officials may lead operations during part or all of this period, but the JFC typically will provide significant supporting capabilities and activities. The joint force may be required to perform limited local governance (i.e., military government), and integrate the efforts of other supporting interagency and multinational partners until legitimate local entities are functioning. The JFC continuously assesses the impact of operations on the ability to transfer authority for remaining requirements to a legitimate civil entity.

(6) **Enable Civil Authority.** Joint force support to legitimate civil governance typically characterizes these actions and activities. The commander provides this support by agreement with the appropriate civil authority. In some cases, especially for operations within the US, the commander provides this support under direction of the civil authority. The purpose is to help the civil authority regain its ability to govern and administer to the services and other needs of the population. The military end state typically is reached during this phase, signaling the end of the joint operation. CCMD involvement with other nations and other government agencies beyond the termination of the joint operation, such as lower-level stability activities and FHA, may be required to achieve national objectives.

Joint Operations Across the Conflict Continuum

For more information on stability activities, refer to JP 3-07, Stability.

c. Some joint operations below the level of large-scale combat will have distinguishable groups of activity. However, activities may be compressed or absent entirely according to the nature of the operation. For example, deployment of forces associated with **seize the initiative** activities may have a deterrent effect sufficient to dissuade an enemy from conducting further operations, returning the OE to a more stable state. Likewise, although FID and NEO may occur as supporting operations to larger combat operations in the OA, they will have no evident dominating activities. Figure V-5 shows a notional successful joint strike, which did not require follow-on operations. Figure V-6 shows a notional FHA operation that required predominantly stabilize and enable civil authority activities.

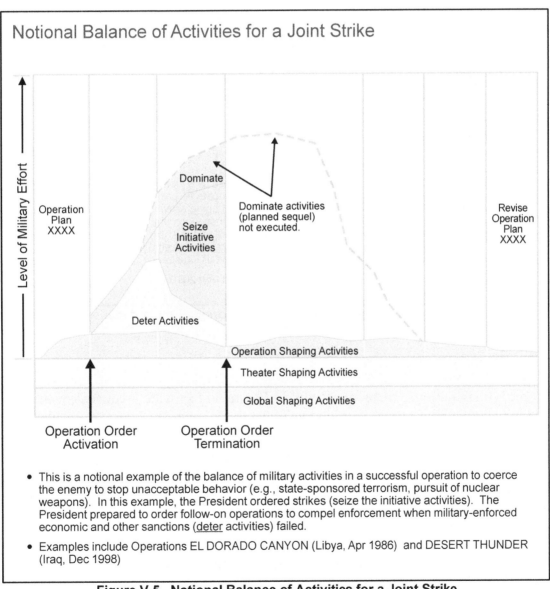

Figure V-5. Notional Balance of Activities for a Joint Strike

V-11

Chapter V

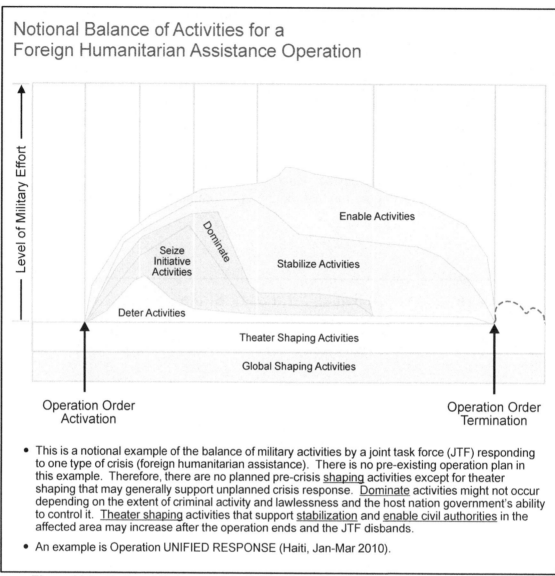

Figure V-6. Notional Balance of Activities for a Foreign Humanitarian Assistance Operation

6. Phasing a Joint Operation

a. The six general groups of activity in Figure V-4 provide a convenient basis for thinking about a joint operation in notional phases, as Figure V-7 depicts. A phase is a definitive stage or period during a joint operation in which a large portion of the forces and capabilities are involved in similar or mutually supporting activities for a common purpose that often is represented by intermediate objectives. Phasing, which can be used in any operation regardless of size, helps the JFC organize large operations by integrating and synchronizing subordinate operations. **Phasing helps JFCs and staffs visualize, plan, and execute the entire operation and define requirements in terms of forces, resources, time, space, and purpose.** It helps them systematically achieve military objectives that cannot be attained all at once by arranging smaller, focused, related

Joint Operations Across the Conflict Continuum

operations in a logical sequence. Phasing also helps commanders mitigate risk in the more dangerous or difficult portions of an operation.

b. Figure V-7 shows one phasing alternative. Actual phases of an operation will vary (e.g., compressed, expanded, or omitted entirely) according to the nature of the operation and the JFC's decisions. For example, UW operations normally use a seven-phase model. During planning, the JFC establishes conditions, objectives, and events for transitioning from one phase to another and plans sequels and branches for potential contingencies. Phases may be conducted sequentially, but some activities from a phase may begin in a previous phase and continue into subsequent phases. The JFC adjusts the phases to exploit opportunities presented by the enemy and operational situation or to react to unforeseen conditions.

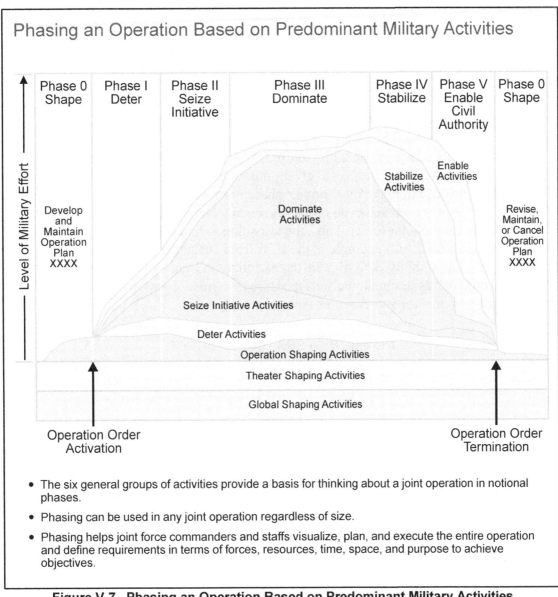

Figure V-7. Phasing an Operation Based on Predominant Military Activities

Chapter V

c. A GCC's theater campaign encompasses all operations and activities for which the GCC is responsible, from relatively benign security cooperation activities through ongoing large-scale combat operations. All six groups of joint operation activities may be present in the GCC's AOR. However, use of the groups of activities for the purpose of phasing applies only to planning and executing individual operations, whether small-scale contingencies or large-scale campaigns that support the GCC's theater campaign. The groups of military activities associated with phases in Figure V-7 can serve as a frame of reference that facilitates common understanding among interagency and multinational partners and supporting commanders of how a JFC intends to execute a specific joint operation as well as progress during execution.

> **The use of groups of activities for the purpose of phasing applies only to planning and executing individual joint operations, not to a GCC's theater campaign or strategy development.**

d. **Transitions**

(1) During execution, a transition marks a change between phases or between the ongoing operations and execution of a branch or sequel. This shift in focus by the joint force often is accompanied by changes in command or support relationships and priorities of effort. Transitions require planning and preparation well before their execution. The activities that predominate during a given phase rarely align with neatly definable breakpoints. The need to move into another phase normally is identified by assessing that a set of objectives has been achieved or that the enemy has acted in a manner that requires a major change for the joint force. Thus, the transition to a new phase is usually driven by events rather than time. An example is the shift from sustained combat operations in the dominate phase to stability activities in the stabilize and enable civil authority phases. Through continuous assessment, the staff measures progress toward planned transitions so that the force prepares for and executes them.

> **COMMON OPERATING PRECEPT**
>
> **Plan for and manage operational transitions over time and space.**

(2) Sometimes, however, the situation facing the JFC will change unexpectedly and without apparent correlation to a planned transition. The JFC may choose to shift operations to address unanticipated critical changes. The JFC must recognize fundamental changes in the situation and respond quickly and smoothly. Failure to do so can cause the joint force to lose momentum, miss important opportunities, experience setbacks, or even fail to accomplish the mission. Conversely, successful transitions enable the joint force to seize the initiative and quickly and efficiently garner favorable results. The JFC should anticipate transformations, as well as plan shifts, during operations.

Refer to JP 5-0, Joint Planning, *for more information on phasing. Refer to JP 3-05.1,* Unconventional Warfare, *for additional information on phasing UW operations. Refer to*

JP 3-20, Security Cooperation, for more information on security cooperation's role in helping set conditions for successful theater operations.

7. The Balance of Offense, Defense, and Stability Activities

a. Combat missions and tasks can vary widely depending on context of the operation and the objective. Most combat operations will require the commander to balance offensive, defensive, and stability activities. This is particularly evident in a campaign or major operation, where combat can occur during several phases and stability activities may occur throughout. Figure V-8 depicts notional proportions of offensive, defensive, and stability activities through the phases of a joint operation.

b. **Offensive and Defensive Operations.** Major operations and campaigns, whether they involve large-scale combat, normally include both offensive and defensive

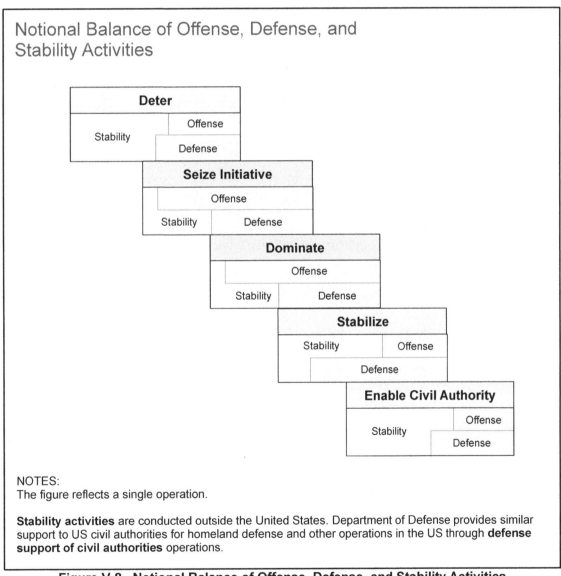

Figure V-8. Notional Balance of Offense, Defense, and Stability Activities

Chapter V

components (e.g., interdiction, maneuver, forcible entry, fire support, countering air and missile threats, DCO, base defense). Although defense may be the stronger form, offense is normally decisive in combat. To achieve military objectives quickly and efficiently, JFCs normally seek the earliest opportunity to conduct decisive offensive operations. Nevertheless, during a sustained offensive, selected elements of the joint force may need to pause, defend, resupply, or reconstitute, while other forces continue the attack. Accordingly, certain defensive measures and protection activities (e.g., OPSEC) are required throughout each joint operation phase. Joint forces at all levels should be capable of rapid transition between offense and defense and vice versa. The relationship between offense and defense, then, is a complementary one. Defensive operations enable JFCs to conduct or prepare for decisive offensive operations.

c. **Stability Activities.** Commanders conduct stability activities to maintain or reestablish a safe and secure environment and provide essential governmental services, emergency infrastructure reconstruction, and humanitarian relief. To achieve objectives and reach the desired military end state, JFCs integrate and synchronize stability activities with offense and defense, as necessary, during the phases of an operation. Stability activities support USG stabilization efforts and contribute to USG initiatives to build partnerships. These initiatives set the conditions to interact with partner, competitor, or adversary leaders, military forces, or relevant populations by developing and presenting information and conducting activities to affect their perceptions, will, behavior, and capabilities. The JFC will likely conduct stability activities in coordination with interorganizational participants and the private sector in support of HN authorities. Stability activities are conducted outside the US. DOD can provide similar support to US civil authorities through DSCA.

For further guidance on stability activities, refer to JP 3-07, Stability.

For further guidance on interorganizational coordination, refer to JP 3-08, Interorganizational Cooperation.

d. **Balance and Simultaneity**

(1) Commanders strive to apply the many dimensions of military power simultaneously across the depth, breadth, and height of the OA. The challenge of balance and simultaneity affects all operations involving combat, particularly campaigns, due to their scope. Consequently, JFCs often concentrate in some areas or on specific functions, and require economy of force in others. However, plans for major operations and campaigns **will normally exhibit a balance between offense and defense and stability activities in various phases. Therefore, planning for stability activities should begin when joint operation planning begins.**

(2) Figure V-8 relates to Figure V-7 and the phasing explanation in paragraph 6, "Phasing a Joint Operation." Figure V-8 illustrates the notional balance between offensive and defensive actions and stability activities as an operation progresses. Since the focus of the CCMD's ongoing theater campaign is on prevention and preparation, any stability activities in the JFC's proposed operational area might continue, and combat (offense and

defense) may be limited or absent. Defensive measures might be limited to providing an increased level of security. A similar balance applies to the deter phase, since the intent is to limit escalation in the OA. A JFC might begin to limit stability activities if an adversary's potential combat actions are imminent. In combat operations, seize the initiative and dominate phases focus on offense and defense. Stability activities are likely restricted to parts of the OA away from immediate combat, or might not occur at all. As the joint force achieves objectives and combat abates, the focus shifts to actions to stabilize and enable civil authority. Stability activities resume and will usually increase in proportion to the decrease in combat.

(3) Planning for the transition from sustained combat operations to assumption of responsibility by civil authority, should begin during plan development and continue during all phases of a joint operation. Planning for redeployment should be considered early and continue throughout the operation and is best accomplished in the same time-phased process in which deployment was accomplished. An unnecessarily narrow focus on planning offensive and defensive operations in the dominate phase may threaten full development of the stabilize and enable civil authority phases and negatively affect joint operation momentum. Even during sustained combat operations the joint force should establish or restore security and control and provide humanitarian relief as areas are occupied, bypassed, or returned to civilian control. Planning for humanitarian assistance should be coordinated through the security cooperation organization and the USAID if it has mission presence, and also shared with the senior development advisor to the CCDR in order to avoid duplication of effort in the HN.

8. Linear and Nonlinear Operations

a. **In linear operations,** each commander directs and sustains combat power toward enemy forces in concert with adjacent units. Linearity refers primarily to the conduct of operations with identified forward lines of own troops (FLOTs). In linear operations, emphasis is placed on maintaining the position of friendly forces in relation to other friendly forces. From this relative positioning of forces, security is enhanced and massing of forces can be facilitated. Also inherent in linear operations is the security of rear areas, especially LOCs between sustaining bases and fighting forces. Protected LOCs, in turn, increase the endurance of joint forces and ensure freedom of action for extended periods. A linear OA organization may be best for some operations or certain phases of an operation. Conditions that favor linear operations include those where US forces lack the information needed to conduct nonlinear operations or are severely outnumbered. Linear operations also are appropriate against a deeply arrayed, echeloned enemy force or when the threat to LOCs reduces friendly force freedom of action. In these circumstances, linear operations allow commanders to concentrate and synchronize combat power more easily. World Wars I and II offer multiple examples of linear operations.

b. **In nonlinear operations,** forces orient on objectives without geographic reference to adjacent forces. Nonlinear operations typically focus on creating specific effects on multiple decisive points. **Nonlinear operations emphasize simultaneous operations along multiple LOOs from selected bases (ashore or afloat).** Simultaneity overwhelms opposing C2 and allows the JFC to retain the initiative. In nonlinear operations, sustaining

Chapter V

functions may depend on sustainment assets moving with forces or aerial delivery. Noncombatants and the fluidity of nonlinear operations require careful judgment in clearing fires, both direct and indirect. Situational awareness, coupled with precision fires, frees commanders to act against multiple objectives. Swift maneuver against several decisive points supported by precise, concentrated fire can induce paralysis and shock among enemy troops and commanders. Nonlinear operations were applied during Operation JUST CAUSE. The joint forces oriented more on their assigned objectives (e.g., destroying an enemy force or seizing and controlling critical terrain or population centers) and less on their geographic relationship to other friendly forces. To protect themselves, individual forces relied more on situational awareness, mobility advantages, and freedom of action than on mass. Nonlinear operations place a premium on the communications, intelligence, mobility, and innovative means for sustainment.

(1) During **nonlinear offensive operations,** attacking forces must focus offensive actions against decisive points, while allocating the minimum essential combat power to defensive operations. Reserves must have a high degree of mobility to respond where needed. JFCs may be required to dedicate combat forces to provide for LOC and base defense. Vulnerability increases as operations extend and attacking forces are exposed over a larger OA. Linkup operations, particularly those involving vertical envelopments, require extensive planning and preparation. The potential for friendly fire incidents increases due to the fluid nature of the nonlinear OA and the changing disposition of attacking and defending forces. The presence of civilians in the OA further complicates operations.

(2) During **nonlinear defensive operations,** defenders focus on destroying enemy forces, even if it means losing physical contact with other friendly units. Successful nonlinear defenses require all friendly commanders to understand the JFCs intent and maintain a common operational picture (COP). Noncontiguous defenses are generally mobile defenses; however, some subordinate units may conduct area defenses to hold key terrain or canalize attackers into engagement areas. Nonlinear defenses place a premium on reconnaissance and surveillance to maintain contact with the enemy, produce relevant information, and develop and maintain a COP. The defending force focuses almost exclusively on defeating the enemy force rather than retaining large areas. Although less challenging than in offensive operations, LOC and sustainment security will still be a test and may require allocation of combat forces to protect LOCs and other high risk functions or bases. The JFC must ensure clear command relationships are established to properly account for the added challenges to base, base cluster, and LOC security.

c. **AOs and Linear/Nonlinear Operations**

(1) **General.** JFCs consider incorporating combinations of contiguous and noncontiguous AOs with linear and nonlinear operations as they conduct operational design. They choose the combination that fits the OE and the purpose of the operation. Association of contiguous and noncontiguous AOs with linear and nonlinear operations creates the four combinations in Figure V-9.

Joint Operations Across the Conflict Continuum

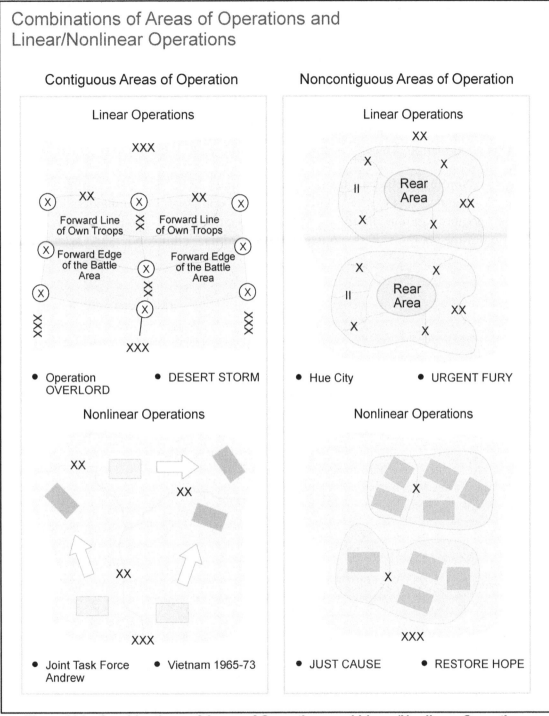

Figure V-9. Combinations of Areas of Operations and Linear/Nonlinear Operations

(2) **Linear Operations in Contiguous AOs.** Linear operations in contiguous AOs (upper left-hand pane in Figure V-9) typify sustained offensive and defensive operations against powerful, echeloned, and symmetrically organized forces. The contiguous areas and continuous FLOT focus combat power and protect sustainment functions.

(3) **Linear Operations in Noncontiguous AOs.** The upper right-hand pane of Figure V-9 depicts a JFC's OA with subordinate component commanders conducting linear operations in noncontiguous AOs. In this case, the JFC retains responsibility for that portion of the OA outside the subordinate commanders' AOs.

(4) **Nonlinear Operations in Contiguous AOs.** The lower left-hand pane in Figure V-9 illustrates the JFC's entire assigned OA divided into subordinate AOs. Subordinate component commanders are conducting nonlinear operations within their AOs. This combination typically is applied in stability activities and DSCA actions.

(5) **Nonlinear Operations in Noncontiguous AOs.** The lower right-hand pane of Figure V-9 depicts a JFC's OA with subordinate component commanders conducting nonlinear operations in noncontiguous AOs. In this case, the JFC retains responsibility for that portion of the operational area outside the subordinate commanders' AOs.

CHAPTER VI
MILITARY ENGAGEMENT, SECURITY COOPERATION, AND DETERRENCE

> *"Building security globally not only assures Allies and partners and builds their capacity but also helps protect the homeland by deterring conflict and increasing stability."*
>
> **Quadrennial Defense Review, 2014**

1. Introduction

a. Military engagement, security cooperation, and deterrence missions, tasks, and actions encompass a wide range of actions where the military instrument of national power is tasked to support other instruments of national power as represented by interagency partners, as well as cooperate with international organizations (e.g., UN, NATO) and other countries to protect and enhance national security interests, deter conflict, and set conditions for future contingency operations. This may also involve domestic operations that include supporting civil authorities. These activities generally occur continuously in all GCCs' AORs regardless of other ongoing joint operations. Military engagement, security cooperation, and deterrence activities usually involve a combination of military forces and capabilities separate from but integrated with the efforts of interorganizational participants. These activities are conducted as part of a CCDR's routine theater or functional campaign plan and country plan objectives and may support deterrence. Because DOS is frequently the major player in these activities, JFCs should maintain a working relationship with the DOS regional bureaus in coordination with the chiefs of the US diplomatic missions and country teams in their area. Commanders and their staffs should establish and maintain dialogue with HN government, multinational partners, and leaders of other organizations pertinent to their operation.

b. Projecting US military force invariably requires extensive use of international waters, international airspace, space, and cyberspace. Military engagement, security cooperation, and deterrence help assure operational access for crisis response and contingency operations despite changing US overseas defense posture and the growth of A2/AD capabilities around the globe. The more a GCC can promote favorable access conditions in advance across the AOR and in potential OAs, the better. Relevant activities include KLEs; security cooperation activities, such as bilateral and multinational exercises to improve multinational operations; missions to train, advise, and equip foreign forces to improve their national ability to contribute to access; negotiations to secure basing and transit rights, establish relationships, and formalize support agreements; the use of grants and contracts to improve relationships with and strengthen PNs; and planning conferences to develop multinational plans.

c. Military engagement, security cooperation, and deterrence activities provide the foundation of the CCDR's theater campaign. The goal is to prevent and deter conflict by keeping adversary activities within a desired state of cooperation and competition. The joint operation model described in Chapter V, "Joint Operations Across the Conflict

Chapter VI

Continuum," has limited application with respect to phasing these activities for normal cooperative and competitive environments. Figure VI-1 shows a notional depiction of activities in an environment of cooperation and competition. DOD forces, as part of larger whole-of-government efforts, conduct operations with partners to prevent, deter, or turn back escalatory activity by adversaries.

(1) Global and theater shaping increases DOD's depth of understanding of an environment, a partner's viewpoint of that environment, and where the US and PN have common interests. This understanding allows the US, through the relationships that have been developed, to shape the OE. These initiatives help advance national security objectives, promote stability, prevent conflicts (or limit their severity), and reduce the risk of employing US military forces in a conflict.

(2) In an environment that is more competitive, tensions increase. A partner's resources can enhance USG understanding of an adversary's capabilities and intent, and expand options against the adversary. In the best case, conflict can be averted or diminished by coordinated USG/PN action.

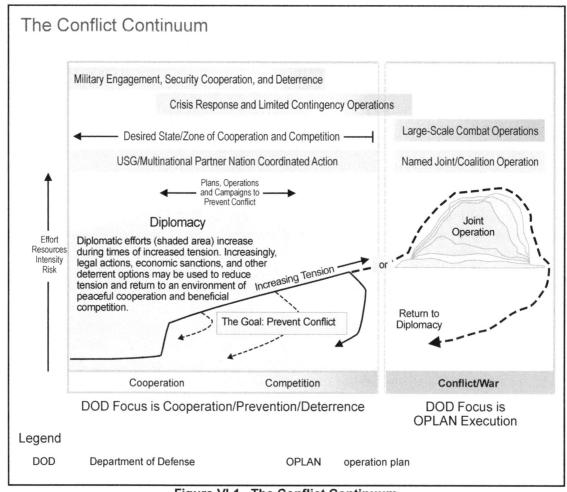

Figure VI-1. The Conflict Continuum

(3) Despite the efforts to prevent or mitigate conflict, an armed conflict may occur. As conditions and objectives become more defined, GCCs may transition to the notional phasing construct for execution of a specific contingency operation as Figure VI-1 depicts. However, time spent "to the left" allows DOD to develop a deeper understanding of the environment to see and act ahead of conflict flashpoints, develop options, and maximize the efficiency of resources.

d. **Military Engagement.** Military engagement is the routine contact and interaction between individuals or elements of the Armed Forces of the United States and those of another nation's armed forces, or foreign and domestic civilian authorities or agencies, to build trust and confidence, share information, coordinate mutual activities, and maintain influence. Military engagement occurs as part of security cooperation, but also extends to interaction with domestic civilian authorities. GCCs seek out partners and communicate with adversaries to discover areas of common interest and tension. This military engagement increases the knowledge base for subsequent decisions and resource allocation. Such military engagements can reduce tensions and may preclude conflict; or, if conflict is unavoidable, allow a more informed USG to enter into it with stronger alliances or coalitions.

e. **Security Cooperation**

(1) Security cooperation involves all DOD interactions with foreign defense establishments to build defense relationships that promote specific US security interests, develop allied and friendly military capabilities for self-defense and multinational operations, and provide US forces with peacetime and contingency access to the HN. The policy on which security cooperation is based resides in Presidential Policy Directive-23, *Security Sector Assistance*. This directive refers to the policies, programs, and activities the US uses to work with foreign partners and help shape their policies and actions in the security sector; help foreign partners build and sustain the capacity and effectiveness of legitimate institutions to provide security, safety, and justice for their people; and, enable foreign partners to contribute to efforts that address common security challenges.

(2) Security cooperation is a key element of global and theater shaping activities and critical aspect of communication synchronization. GCCs shape their AORs through security cooperation and stability activities by continually employing military forces to complement and support other instruments of national power that typically provide development assistance or humanitarian assistance to PNs. The GCC's TCP provides a framework within which CCMDs conduct cooperative security cooperation activities and development with PNs. Ideally, security cooperation activities mitigate the causes of a potential crisis before a situation deteriorates and requires US military intervention. Security assistance and security force assistance (SFA) normally provide some of the means for security cooperation activities.

Refer to JP 3-20, Security Cooperation, *for more information on security cooperation.*

Chapter VI

f. **Deterrence**

(1) Deterrence prevents adversary action through the presentation of a credible threat of unacceptable counteraction and belief that the cost of the action outweighs the perceived benefits. The nature of deterrent options varies according to the nature of the adversary (e.g., traditional or irregular, state or non-state), the adversary's actions, US national objectives, and other factors. Deterrence stems from an adversary's belief that the opponent's actions have created or can create an unacceptable risk to the adversary's achievement of objectives (i.e., the contemplated action cannot succeed or the costs are too high). Thus, a potential aggressor chooses not to act for fear of failure, risk, or consequences. Ideally, deterrent forces should be able to conduct decisive operations immediately. However, if available forces lack the combat power to conduct decisive operations, they conduct defensive operations while additional forces deploy. Effective deterrence requires a TCP and a coordinated CCS effort that emphasize security cooperation activities with PNs that support US interests, DOD force posture planning, and contingency plans that prove the willingness of the US to employ forces in defense of its interests. Various joint operations (e.g., show of force and enforcement of sanctions) support deterrence by demonstrating national resolve and willingness to use force when necessary. Other TCP actions that help maintain or set the CCDR's desired conditions support deterrence by enhancing a climate of peaceful cooperation and FHA, thus promoting stability. Joint actions such as antiterrorism, DOD support to CD operations, show of force operations, and arms control are applied to meet military engagement, security cooperation, and deterrence objectives.

(2) Sustained presence contributes to deterrence and promotes a secure environment in which diplomatic, economic, and informational programs designed to reduce the causes of instability can perform as designed. Presence can take the form of forward basing, forward deploying, or pre-positioning assets. Forward presence activities demonstrate our commitment, lend credibility to our alliances, enhance regional stability, and provide a crisis response capability while promoting US influence and access. Joint force presence often keeps unstable situations from escalating into larger conflicts. The sustained presence of strong, capable forces is the most visible sign of US commitment to allies and adversaries alike. However, if sustained forward presence fails to deter an adversary, committed forces must be agile enough to transition rapidly to combat operations. In addition to forces stationed overseas and afloat, forward presence involves periodic rotational deployments and redeployments, access and storage agreements, multinational exercises, port visits, foreign military training, foreign community support, and both military-to-military and military-to-civilian contacts. Given their location and knowledge of the region, forward presence forces could be the first that a CCDR commits to respond to a crisis. At the same time, commanders must consider adversaries' perceptions of forward presence and deterrent acts. While the deterrent intent of an act may be clear to the actor, adversaries may perceive such acts as hostile and respond in kind. Carefully calculated forward presence, clearly signaled to adversaries as non-aggressive, can prevent escalation, but poorly calculated or poorly signaled increases in forward presence can lead to counter-productive countermeasures and escalation. CCDRs use their situational understanding of the OE to advise political leaders regarding possible reactions to any decision about deploying forces forward as a deterrent.

2. Typical Operations and Activities

a. **Military Engagement Activities.** Numerous routine missions (e.g., security cooperation) and continuing operations or tasks (e.g., freedom of navigation) occur globally on a continuing basis under the general heading of military engagement. These activities build strong relationships with partners, increase regional awareness and knowledge of a PN's capabilities and capacity, and can be used to influence events in a desirable direction. Military engagement activities can also increase understanding of an adversary's capabilities, capacity, and intentions and can provide forewarning of undesirable events. In some cases, what begins as a military engagement activity (e.g., limited support to COIN through a security assistance program) can expand to a limited contingency operation or even a major operation when the President commits US forces. Military engagement activities are generally governed by various directives and agreements and do not require a joint OPLAN or OPORD for execution.

b. **Emergency Preparedness.** Emergency preparedness consists of measures taken in advance of an emergency to reduce the loss of life and property and to protect a nation's institutions from all types of hazards through a comprehensive emergency management program of preparedness, mitigation, response, and recovery. At the strategic level, emergency preparedness encompasses those planning activities, such as continuity of operations and continuity of government, undertaken to ensure DOD processes, procedures, and resources are in place to support the President and SecDef in a designated national security emergency.

(1) Continuity of operations ensures continuous conduct of functions, tasks, or duties necessary to accomplish a military action or mission supporting the national strategy. Continuity of operations includes the functions and duties of the commander, as well as the supporting functions and duties performed by the staff and others under the authority and direction of the commander. If the President directs, DOD may be tasked with additional missions relating to emergency preparedness.

(2) Continuity of government involves a coordinated effort within each USG branch (executive, legislative, and judicial) to ensure the capability to continue minimum essential functions and responsibilities during a catastrophic emergency.

c. **Arms control, nonproliferation, and disarmament** are not synonymous. The following are examples of US military personnel involvement in arms control, nonproliferation, and disarmament activities: verifying an arms control treaty; seizing and securing WMD; escorting authorized deliveries of weapons and other materials (e.g., enriched uranium) to preclude loss or unauthorized use of these assets; conducting and hosting site inspections; participating in military data exchanges; implementing armament reductions; or dismantling, destroying, or disposing of weapons and hazardous material.

(1) Arms control agreements refer to the written or unwritten embodiment of the acceptance of one or more arms control measure by two or more nations. Arms control refers to mutually agreed upon and verifiable restraints between states on the research, manufacture, or levels of, and/or locations of, deployment of troops and weapon systems.

Chapter VI

Arms control may be used by states to restrain military modernization of another party to the agreement, to entitle a party to achieve at least military parity with another party, to free up hard currency for other priorities, to redirect military modernization efforts so as to take advantage of technological advances in new weapon system areas, to facilitate sources and methods of intelligence gathering, or to enable states to negate first strike capabilities.

(2) Nonproliferation includes use of military capabilities in conjunction with a whole-of-government effort, and within a state's legal authorities, to deter and prevent the acquisition of WMD by dissuading or impeding access to or distribution of sensitive technologies, material, and expertise by and between state and non-state actors of concern. Usually sanctions are established by UN Security Council resolutions. However, states may view the need to unilaterally, or in concert, sanction certain military equipment as a necessity of their national interests. Aggressive military force is rarely employed in nonproliferation military operations or activities (e.g., except in self-defense).

(3) Disarmament is the reduction of a military establishment (e.g., the number of weapons and troops maintained by a state) to some level set by international agreement. Although disarmament always involves the reduction of military forces or weapons, arms control does not. In fact, arms control agreements sometimes allow for the increase of weapons by one or more parties to a treaty. Disarmament requires a high degree of trust (permissive OE), and disarmament operations are unlikely between hostile nations.

d. **Combating Terrorism.** Combating terrorism involves actions to oppose terrorism from all threats. It encompasses antiterrorism—defensive measures taken to reduce vulnerability to terrorist acts—and CT—offensive measures to prevent, deter, preempt, and respond to terrorism.

(1) **Antiterrorism** involves defensive measures to reduce the vulnerability of individuals and property to terrorist acts, to include limited response and containment by local military and civilian forces. Antiterrorism programs form the foundation to combat terrorism. The USG may provide antiterrorism assistance to foreign countries under Title 22, USC (under Antiterrorism Assistance).

For further guidance on antiterrorism, refer to JP 3-07.2, Antiterrorism.

(2) **CT.** CT is primarily a special operations core activity and consists of activities and operations taken to neutralize terrorists and their organizations and networks in order to render them incapable of using violence to instill fear and coerce governments or societies to achieve their goals. CT includes direct actions against terrorist networks and indirect actions to influence and render global and regional environments inhospitable to terrorist networks. Normally, CT operations require specially trained and equipped personnel capable of swift and effective action. **CT is often associated with IW.** However, terrorists also operate during large-scale traditional combat, forcing commanders to integrate CT with these operations. Enemies using irregular methods often will use terrorist tactics to wage prolonged operations to break the opponent's will and influence relevant populations. At the same time, terrorists and insurgents also seek to bolster their own legitimacy and credibility with those same populations. Therefore, CT efforts should

> **JOINT TASK FORCE-NORTH**
>
> An example of Department of Defense support to counterdrug operations was the establishment of Joint Task Force (JTF)-6 in 1989. Its mission originally focused exclusively along the Southwest border of the US. A succession of National Defense Authorization Acts expanded the JTF-6 charter by adding specific mission tasks for the organization. In 1995, the JTF-6 area of operations expanded to include the continental US. In June 2004, JTF-6 was officially renamed JTF-North and its mission was expanded to include providing support to federal law enforcement agencies in countering transnational threats.
>
> Mission: JTF-North supports drug law enforcement agencies in the conduct of counterdrug/counter narcoterrorism operations in the US Northern Command area of responsibility to disrupt transnational criminal organizations and deter their freedom of action in order to protect the homeland.
>
> **Various Sources**

include all instruments of national power to undermine enemy power, will, credibility, and legitimacy, thereby diminishing its ability to influence the relevant population.

For further details concerning CT and special operations, refer to JP 3-26, Counterterrorism, *and JP 3-05,* Special Operations. *For US policy on CT, refer to the* National Strategy for Combating Terrorism.

   e. **Support to CD Operations.** DOD supports federal, state, and local LEAs in their effort to **disrupt the transport and/or transfer of illegal drugs into the US.** Specific DOD authorities that pertain to a CD are contained in Title 10, USC, Sections 124 and 371-382. Discussion of similar authorities is discussed in DODI 3025.21, *Defense Support of Civilian Law Enforcement Agencies.*

For additional guidance on CD operations, refer to JP 3-07.4, Counterdrug Operations.

   f. **Sanction enforcement** is any operation that employs coercive measures to control the movement of designated items into or out of a nation or specified area. **Maritime interception operations** are efforts to monitor, query, and board merchant vessels in international waters to enforce sanctions against other nations such as those in support of UN Security Council resolutions and/or prevent the transport of restricted goods. These operations serve both strategic and military purposes. The strategic objective is to compel a country or group to conform to the objectives of the initiating body, while the military objective focuses on establishing a selective barrier that allows only authorized goods to enter or exit. Depending on the geography, **sanction enforcement normally involves some combination of air and surface forces.** Assigned forces should be capable of **complementary mutual support** and **full communications interoperability.**

Chapter VI

> **OVERFLIGHT AND FREEDOM OF NAVIGATION OPERATIONS**
>
> **The Berlin air corridors, established between 1948 and 1990, which allowed air access to West Berlin, were set up to maintain international airspace to an "air-locked" geographical area. When Soviet forces disrupted ground passage to Berlin, the US asserted its rights to utilize these air corridors during the Berlin airlift. The ATTAIN DOCUMENT series of operations against Libya in 1986 were freedom of navigation operations, both air and sea, in the Gulf of Sidra, a recognized international body of water over which Libya illegally claimed sovereignty.**
>
> **Various Sources**

g. **Enforcement of Exclusion Zones.** A sanctioning body establishes an exclusion zone to **prohibit specified activities in a specific geographic area.** Exclusion zones usually are imposed due to **breaches of international standards of human rights or flagrant violations of international law** regarding the conduct of states. Situations that may warrant such action include persecution of civil populations by a government and efforts by a hostile nation to acquire territory by force. Exclusion zones can be established in the air (no-fly zones), sea (maritime), or on land (no-drive zones). An exclusion zone's purpose may be to persuade nations or groups to modify their behavior to meet the desires of the sanctioning body or face continued imposition of sanctions or threat or use of force. **Such measures are usually imposed by the UN or another international body** of which the US is a member, although they may be imposed unilaterally by the US (e.g., Operation SOUTHERN WATCH in Iraq, initiated in August 1992, and Operation DENY FLIGHT in Bosnia, from March 1993 to December 1995).

h. **Freedom of Navigation and Overflight.** Freedom of navigation operations are conducted to protect US navigation, overflight, and related interests on, under, and over the seas, against excessive maritime claims. Freedom of navigation is a sovereign right accorded by international law.

(1) International law has long recognized that a coastal state may exercise jurisdiction and control within its territorial sea in the same manner that it can exercise sovereignty over its own land territory. International law accords the right of "innocent" passage to ships of other nations through a state's territorial waters. Passage is "innocent" as long as it is not prejudicial to the peace, good order, or security of the coastal state. The high seas are free for reasonable use of all states.

(2) Freedom of navigation by aircraft through international airspace is a well-established principle of international law. Aircraft threatened by nations or groups through the extension of airspace control zones outside the established international norms will result in a measured legal response, appropriate to the situation. The **International Civil Aviation Organization,** a specialized agency of the UN, codifies the principles and techniques of international air navigation and fosters the planning and development of international air transport to ensure safe and orderly use of international airspace.

i. **Foreign assistance** is **civil or military assistance rendered to a nation by the USG** within that nation's territory based on agreements mutually concluded between the US and that nation (e.g., Operation PROMOTE LIBERTY, in 1990, following Operation JUST CAUSE in Panama). Foreign assistance supports the HN by promoting sustainable development and growth of responsive institutions. **The goal is to promote long-term regional stability.** Foreign assistance programs include security assistance, development assistance, and humanitarian assistance, and can support FID and stability activities. To be effective, foreign assistance should include collaborative planning among the JFC, DOS, USAID, the chief of mission, the country team in the HN, HN authorities, and any supporting international organization or NGO. Normally, DOD foreign assistance activities in an HN are integrated into and support objectives of the chief of mission's integrated country strategy, which is consolidated in the TCP and the country-specific security cooperation section/country plans that are nested within the TCP.

j. Security assistance is a group of programs by which the US provides defense articles, military training, and other defense-related services to foreign nations by grant, loan, credit, or cash sales in furtherance of national policies and objectives. These programs are funded and authorized by DOS to be administered by DOD and the Defense Security Cooperation Agency. They are an element of security cooperation. Some **examples of US security assistance programs** are the Foreign Military Sales Program, the Foreign Military Financing Program, the International Military Education and Training Program, the Economic Support Fund, and commercial sales licensed under the Arms Export Control Act. **Security assistance surges can accelerate release of equipment, supplies, or services** when an allied or friendly nation faces an imminent military threat. Security assistance surges are primarily **military** and provide additional combat systems (e.g., weapons and equipment) or supplies, but may include the full range of security assistance, to include financial and training support.

k. **SFA.** SFA is DOD's contribution to unified action by the USG to support the development of the capacity and capability of foreign security forces (FSF) and their supporting institutions, to achieve objectives shared by the USG. **SFA is conducted with and through FSF.** The US military conducts activities to enhance the capabilities and capacities of a PN (or regional security organization) by providing training, equipment, advice, and assistance to those FSF organized in national ministry of defense (or equivalent regional military or paramilitary forces). Other USG departments and agencies focus forces assigned to other ministries (or their equivalents) such as interior, justice, or intelligence services.

For further information about security cooperation, security assistance, and SFA, refer to JP 3-20, Security Cooperation.

l. **FID** encompasses participation by civilian and military agencies of a government in the action programs taken by another government or other designated organization to free and protect its society from subversion, lawlessness, insurgency, terrorism, and other threats to its security. USG support to FID can include diplomatic, economic, informational, and military development assistance to HN security sector and collaborative planning with multinational and HN authorities to anticipate, preclude, and counter those

Chapter VI

threats. US military involvement in FID has traditionally focused on helping a nation defeat an organized movement attempting to violently overthrow its lawful government. US FID programs may address other threats to the stability of an HN, such as civil disorder, illicit weapons, drug and human trafficking, and terrorism. While FID is a legislatively mandated special operations core activity, conventional forces also contain and employ organic capabilities to conduct SFA activities for FID.

For further guidance on FID, refer to JP 3-22, Foreign Internal Defense. *For further guidance on SOF involvement in FID, refer to JP 3-05,* Special Operations.

m. **Humanitarian assistance** programs are governed by Title 10, USC, Section 401. This assistance may be provided in conjunction with military operations and exercises, but must fulfill unit training requirements that incidentally create humanitarian benefit to the local populace. In contrast to emergency relief conducted under FHA operations, humanitarian and civic assistance programs generally encompass planned activities in the following categories:

(1) **Medical, dental, and veterinary care** provided in rural or underserved areas of a country.

(2) Construction and repair of basic **surface transportation systems.**

(3) **Well drilling** and construction of basic **sanitation facilities.**

(4) Rudimentary construction and repair of **public facilities** such as schools, health and welfare clinics, and other nongovernmental buildings.

n. **Protection of Shipping.** When necessary, **US forces provide protection** of US flag vessels, US citizens (whether embarked in US or foreign vessels), and US property **against unlawful violence in and over international waters** (such as Operation EARNEST WILL, in which Kuwaiti ships were reflagged under the US flag in 1987). This protection may be extended to foreign flag vessels under international law and with the consent of the flag state. Actions to protect shipping include **coastal sea control, harbor defense, port security, countermine operations,** and **environmental defense,** in addition to operations on the high seas. Protection of shipping, which is a critical element in the fight against piracy, requires the coordinated employment of surface, air, space, and subsurface units, sensors, and weapons, as well as a command structure both ashore and afloat and a logistic base. Protection of shipping may require a combination of operations to be successful. These actions can include area operations, escort duties, mine countermeasures, and environmental defense missions.

o. **Show of force operations** are designed to demonstrate US resolve. They involve the **appearance of a credible military force** in an attempt to defuse a situation that, if allowed to continue, may be detrimental to US interests. These operations also underscore US commitment to our multinational partners.

(1) The US deploys forces abroad to **lend credibility** to its promises and commitments, **increase its regional influence,** and **demonstrate its resolve to use**

> **SHOW OF FORCE IN THE PHILIPPINES**
>
> Operation Joint Task Force-PHILIPPINES was conducted by US forces in 1989 in support of President Aquino during a coup attempt against the Philippine government. During this operation, a large special operations force was formed, fighter aircraft patrolled above rebel air bases, and two aircraft carriers were positioned off the coastline of the Philippines.
>
> Various Sources

military force if necessary. In addition, SecDef orders a show of force to bolster and reassure friends and allies. Show of force operations are military in nature but often serve both diplomatic and military purposes. These operations may influence other governments or politico-military organizations to refrain from belligerent acts.

(2) **Diplomatic concerns dominate a show of force operation,** and as such, military forces often are under significant legal and diplomatic constraints and restraints. The military force coordinates its operations with the country teams affected. A show of force can involve a wide range of military forces including joint US or multinational forces. Often, bilateral or multinational training and exercises are scheduled to demonstrate strength and resolve. Forces conducting a show of force operation are also capable of FDOs, FROs, and transitioning to crisis response or limited contingency activities.

p. **Support to Insurgency**

(1) An insurgency is the organized use of subversion and violence to seize, nullify, or challenge political control of a region. Insurgency can also refer to the group itself. Insurgents use a mixture of political, economic, informational, and combat actions to achieve political aims. Insurgency is a protracted politico-military struggle designed to weaken the control and legitimacy of an established government, an interim governing body, or a peace process, while simultaneously increasing insurgent control and legitimacy. Legitimacy is the central issue in an insurgency.

(2) The US may support insurgencies that oppose oppressive regimes. The US coordinates this support with its friends and allies. US military support is typically through **UW,** which includes activities to enable a resistance movement or insurgency to coerce, disrupt, or overthrow a government or occupying power by operating with an underground, auxiliary, and guerrilla force in a denied area. Special forces are well-suited to conduct UW and provide this support. Conventional forces have functional specialties they may contribute to the mission. US forces may provide logistic and training support, as they did for the Mujahidin resistance in Afghanistan during the Soviet occupation in the 1980s. In certain circumstances, the US can provide direct combat support, such as support to the French Resistance in World War II, the Afghanistan Northern Alliance to remove the Taliban in 2001-2002, and for NATO's liberation of Kosovo in 1999.

For further guidance on support to insurgency, refer to JP 3-05.1, Unconventional Warfare.

Chapter VI

q. **COIN** operations include civilian and military efforts designed to support a government in the military, paramilitary, political, economic, psychological, and civic actions it undertakes to defeat insurgency and address its root causes. Insurgents use irregular forms of warfare to undermine their enemies' legitimacy and credibility. Ultimately, insurgents seek to isolate their enemies from the relevant populations and their external supporters, physically as well as psychologically. At the same time, they also seek to bolster their own legitimacy and credibility to exercise authority over that same population. COIN operations often include security assistance programs such as foreign military sales, foreign military financing, and international military education and training. Such support may also include FID and SFA. In some cases, US COIN operations can be much more extensive and involve joint force limited contingency or major operations.

For further guidance on support to COIN, refer to JP 3-24, Counterinsurgency Operations, *and JP 3-22,* Foreign Internal Defense.

3. Other Considerations

a. **Interagency, International, and Nongovernmental Organizations and HN Coordination.** JFCs will work with interorganizational and HN authorities to plan and conduct military engagement, security cooperation, and deterrence operations and activities. Liaison organizations such as a JIACG can promote interaction and cooperation among diverse agencies. Consensus building improves each agency's understanding of the capabilities and limitations, as well as any constraints, of partner agencies. Establishing an atmosphere of trust and cooperation promotes unity of effort to accomplish USG objectives.

For further discussion on interorganizational coordination, refer to JP 3-08, Interorganizational Cooperation.

b. **Information Sharing.** NGOs and international organizations, by the nature of what they do, become familiar with the culture, language, sensitivities, and status of the populace, as well as the infrastructure in a region. This information is valuable to commanders and staffs who may not have physical access or the most current information. NGOs and international organizations may also need information from commanders and staffs concerning security issues. However, these organizations hold neutrality as a fundamental principle. Many NGOs and international organizations will resist being used as sources of intelligence, and they may be hesitant to associate with the military. Discrete coordination can sometimes alleviate these concerns. JFCs may elect to establish mechanisms like a CMOC, or a similar organization, to coordinate activities and facilitate information sharing. International organizations and NGOs are more likely to participate if they perceive that mutual sharing of information aids their work and is not a threat to their neutrality. USAID, when it has a mission presence in country, usually has the strongest network of contacts and information on international organizations, NGOs, and local partners and should be consulted. USAID missions are required to share their Country Development Cooperation Strategies with CCMDs, and conversely, CCMDs are encouraged to share their TCPs with USAID missions in their AOR to enhance information sharing.

c. **Cultural Awareness.** Military support and operations intended to support an HN should be built on in-depth understanding of the HN's cultural, social, economic, and political realities. The JFC may augment Service-language and cultural awareness training and tailor supplemental training to the JOA and mission. Also, intelligence products and military engagement actions continuously update cultural, social, economic, and political information. The beliefs, perceptions, lifestyles, and economic underpinnings of the society, among other considerations, influence the OE and will affect planning and execution. Further, it is important to monitor perceptions and reactions of populations in the areas of influence and area of interest, as these factors also affect current and future operations, activities, and planning.

(1) Security cooperation activities will likely impact countries throughout a region. Traditional rivalries among neighboring states and hostility toward the US may be factors. For example, US assistance to a nation with long-standing rivals in the area may be perceived by these rivals as upsetting the regional balance of power. While such factors do not dictate US policy, they should be carefully evaluated and considered prior to military operations.

(2) Multinational efforts to enhance stability, foster development, and prevent conflict, combined with contingency operations to contain non-state regional actors, require JFCs and staff to plan for sufficient permanent and supplemental (surge) interpreter, translator, and cultural analysis capabilities.

Intentionally Blank

CHAPTER VII
CRISIS RESPONSE AND LIMITED CONTINGENCY OPERATIONS

> *"If we are to retain... a choice other than nuclear holocaust or retreat, we must be ready to fight a limited war for a protracted period of time anywhere in the world."*
>
> **John F. Kennedy: Message on the Budget, Fiscal Year 1963**

1. Introduction

Crisis response and limited contingency operations typically are focused in scope and scale and conducted to achieve a very specific strategic or operational-level objective in an OA. They may be conducted as a stand-alone response to a crisis (e.g., NEO) or executed as an element of a larger, more complex operation. Joint forces conduct crisis response and limited contingency operations to achieve operational and, sometimes, strategic objectives.

2. Crisis Response and Limited Contingency Operations

a. CCDRs plan for various situations that require military operations in response to natural disasters, terrorists, subversives, or other contingencies and crises as directed by appropriate authority. The level of complexity, duration, and resources depends on the circumstances. Limited contingency operations ensure the safety of US citizens and US interests while maintaining and improving the ability to operate with multinational partners to deter hostile ambitions of potential aggressors (e.g., **JTF SHINING HOPE** in the spring of 1999 to support refugee humanitarian relief for hundreds of thousands of ethnic Albanians fleeing their homes in Kosovo). Many of these operations involve a combination of military forces and capabilities operating in close cooperation with interorganizational participants. APEX integrates crisis action and deliberate planning into one unified construct to facilitate unity of effort and transition from planning to execution. Planning functions can be performed in series over a period of time or they can be compressed, performed in parallel, or truncated as appropriate.

b. **Initial Response.** When crises develop and the President directs, CCDRs respond. If the crisis revolves around external threats to a regional partner, CCDRs employ joint forces to deter aggression and signal US commitment (e.g., deploying joint forces to train in Kuwait). If the crisis is caused by an internal conflict that threatens regional stability, US forces may intervene to restore or guarantee stability (e.g., Operation RESTORE DEMOCRACY, the 1994 intervention in Haiti). If the crisis is within US territory (e.g., natural or man-made disaster, deliberate attack), US joint forces will conduct DSCA and HD operations as directed by the President and SecDef. Prompt deployment of sufficient forces in the initial phase of a crisis can preclude the need to deploy larger forces later. Effective early intervention can also deny an adversary time to set conditions in their favor, achieve destabilizing objectives, or mitigate the effects of a natural or man-made disaster. Deploying a credible force rapidly is one step in deterring or blocking aggression. However, deployment alone will not guarantee success. Achieving successful deterrence involves convincing the adversary that the deployed force is able to conduct decisive

Chapter VII

operations and the national leadership is willing to employ that force and to deploy more forces if necessary.

c. **Strategic Aspects.** Two important aspects about crisis response and foreign limited contingency operations stand out. **First, understanding the strategic objective helps avoid actions that may have adverse diplomatic or political effects.** It is not uncommon in some operations, such as peacekeeping, for junior leaders to make decisions that have significant strategic implications. **Second, commanders should remain aware of changes not only in the operational situation, but also in strategic objectives that may warrant a change in military operations.** These changes may not always be obvious. Therefore, commanders must strive to detect subtle changes, which may eventually lead to disconnects between national objectives and military operations. Failure to recognize changes in national objectives early may lead to ineffective or counterproductive military operations.

d. **Economy of Force.** The strategic environment requires the US to maintain and prepare joint forces for crisis response and limited contingency operations simultaneously with other operations, preferably in concert with allies and/or PNs when appropriate. This approach recognizes that these operations will vary in duration, frequency, intensity, and the number of personnel required. The burden of many crisis response and limited contingency operations may lend themselves to using small elements like SOF in coordination with allied nations or PNs. Initially, SOF may take the lead of these operations as an economy of force measure to enable major operations and campaigns with conventional focus to progress more effectively.

3. **Typical Operations**

a. **NEOs** are operations directed by DOS or other appropriate authority, in conjunction with DOD, whereby noncombatants are evacuated from locations within foreign countries to safe havens designated by DOS when their lives are endangered by war, civil unrest, or natural disaster. Although principally conducted to evacuate US citizens, NEOs may also include citizens from the HN, as well as citizens from other countries. Pursuant to Executive Order 12656, *Assignment of Emergency Preparedness Responsibilities,* DOS is responsible for the protection and evacuation of US citizens abroad and for safeguarding their property. This order also directs DOD to advise and assist DOS to prepare and implement plans for the evacuation of US citizens. The US ambassador, or chief of the diplomatic mission, prepares the emergency action plans that address the military evacuation of US citizens and designated foreign nationals from a foreign country. The GCC conducts military operations to assist in the implementation of emergency action plans as directed by SecDef.

For additional guidance on NEOs, refer to JP 3-68, Noncombatant Evacuation Operations.

b. **PO.** PO are multiagency and multinational operations involving all instruments of national power—including international humanitarian and reconstruction efforts and military missions—to contain conflict, restore the peace, and shape the environment to support reconciliation and rebuilding and facilitate the transition to legitimate governance.

OPERATION ATLAS RESPONSE

In the early part of February 2000, Cyclone Connie drenched the Southern Africa region with over 40 inches of rain causing many rivers in the region to overflow and flood populated areas. US European Command sent a humanitarian assistance survey team (HAST) to get "eyes on the ground." Just as the effects of Connie were lessening and the HAST was preparing to head home, Cyclone Leon-Eline hit Madagascar. The storm pushed further inland and rain fell in Zimbabwe, adding to reservoirs that were already full. This forced the release of water from reservoirs, causing even more flooding. Mozambique was the country with the greatest needs in the region. Consequently, between 18 February and 1 April 2000, Joint Task Force (JTF)-ATLAS RESPONSE, under the command of Major General Joseph H. Wehrle, Jr., US Air Force, was sent to aid the people of Mozambique, South Africa, Botswana, Zimbabwe, and Zambia.

The joint force commander established a small, main headquarters in Maputo, Mozambique, to be near the US Ambassador. The majority of forces and staff resided at Air Force Base Hoedspruit, South Africa. Eventually, a small contingent of forces would deploy to Beira, Mozambique, to work at a supply distribution hub. The primary predeployment tasks of the JTF: 1) Search and rescue (SAR), 2) Coordination and synchronization of relief efforts, and 3) Relief supply distribution. Upon arrival, the JTF discovered SAR efforts were essentially complete and a fourth key task became the conduct aerial assessment of the lines of communications. This fourth task was important because it was also a key indicator in the exit strategy.

During the brief time of the operation, the JTF's aircraft carried a total of 714.3 short tons of intergovernmental organization (IGO)/nongovernmental organization (NGO) cargo, most of it for direct support of the local population. Helicopters and C-130s also moved 511 non-US passengers. The majority were medics or aid workers carried on special operations forces HH-60s bringing immediate relief to populations cut off from the rest of the world.

Operation ATLAS RESPONSE was a political and military success. Not only was humanitarian aid provided to the people of Mozambique, but good relations with South African military and many IGOs and NGOs were forged.

SOURCE: Derived from Dr. Robert Sly's, "ATLAS RESPONSE Study," Third Air Force History Office, 2000

For the Armed Forces of the United States, PO encompass PKO, predominantly military PEO, predominantly diplomatic PB actions, PM processes, and conflict prevention. PO are conducted in conjunction with the various diplomatic activities and humanitarian efforts necessary to secure a negotiated truce and resolve the conflict. PO are tailored to each situation and may be conducted in support of diplomatic activities before, during, or after conflict. PO support national/multinational strategic objectives. Military support

Chapter VII

improves the chances for success in the peace process by lending credibility to diplomatic actions and demonstrating resolve to achieve viable political settlements.

For additional guidance on PO, refer to JP 3-07.3, Peace Operations.

c. **FHA.** FHA operations relieve or reduce human suffering, disease, hunger, or privation in countries outside the US. These operations are different from foreign assistance primarily because they occur on short notice as a contingency operation to provide aid in specific crises or similar events rather than as more deliberate foreign assistance programs to promote long-term stability. DOS or the chief of mission in country is responsible for confirming the HN's declaration of a foreign disaster or situation that requires FHA. FHA provided by US forces is generally limited in scope and duration; it is intended to supplement or complement efforts of HN civil authorities or agencies with the primary responsibility for providing assistance. DOD provides assistance when the need for relief is gravely urgent and when the humanitarian emergency dwarfs the ability of normal relief agencies to effectively respond.

For further guidance on FHA operations, refer to JP 3-29, Foreign Humanitarian Assistance.

d. Recovery operations may be conducted to search for, locate, identify, recover, and return isolated personnel, sensitive equipment, items critical to national security, or human remains (e.g., JTF **Full Accounting,** which had the mission to achieve the fullest possible accounting of Americans listed as missing or prisoners of war from all past wars and conflicts). Regardless of the recovery purpose, each type of recovery operation is generally a sophisticated activity requiring detailed planning in order to execute. Recovery operations may be clandestine, covert, or overt depending on whether the OE is hostile, uncertain, or permissive.

e. **Strikes and Raids**

(1) **Strikes** are attacks conducted to damage or destroy an objective or a capability. Strikes may be used to punish offending nations or groups, uphold international law, or prevent those nations or groups from launching their own attacks (e.g., Operation EL DORADO CANYON conducted against Libya in 1986, in response to the terrorist bombing of US Service members in Berlin). Although often tactical in nature with respect to the ways and means used and duration of the operation, strikes can achieve strategic objectives as did the strike against Libya.

(2) **Raids** are operations to temporarily seize an area, usually through forcible entry, in order to secure information, confuse an enemy, capture personnel or equipment, or destroy an objective or capability (e.g., Operation RHINO, a raid led by US SOF elements on several Taliban targets in and around Kandahar, Afghanistan, in October 2001). Raids end with a planned withdrawal upon completion of the assigned mission.

f. **HD and DSCA.** Security and defense of the US homeland is the USG's top responsibility and is conducted as a continuous, cooperative effort among all federal

> **OPERATION EL DORADO CANYON**
>
> The strike was designed to hit directly at the heart of Libyan leader Muammar Gaddafi's ability to export terrorism with the belief that such a preemptive strike would provide him "incentives and reasons to alter his criminal behavior." The final targets were selected at the National Security Council level "within the circle of the President's advisors." Ultimately, five targets were selected. All except one of the targets were chosen because of their direct connection to terrorist activity. The single exception was the Benina military airfield which based Libyan fighter aircraft. This target was hit to preempt Libyan interceptors from taking off and attacking the incoming US bombers.
>
> The actual combat commenced at 0200 (local Libyan time) and lasted less than 12 minutes, resulting in the dropping of 60 tons of munitions. Navy A-6 Intruders were assigned the two targets in the Benghazi area, and the Air Force F-111s hit the other three targets in the vicinity of Tripoli. Resistance outside the immediate area of attack was nonexistent, and Libyan air defense aircraft never launched. One F-111 strike aircraft was lost during the strike.
>
> **Various Sources**

agencies, as well as state, tribal, and local government. Military operations inside the US and its territories, though limited in many respects, are conducted to accomplish two missions—HD and DSCA.

(1) **HD.** HD is the protection of US sovereignty, territory, domestic population, and critical defense infrastructure against external threats and aggression **or other threats as directed by the President. DOD is the federal agency with lead responsibility, supported by other agencies, to defend against external threats and aggression.** However, against internal threats DOD may be in support of another USG department or agency. **When ordered to conduct HD operations within US territory, DOD will coordinate closely with other government agencies.** Consistent with laws and policy, the Services will provide capabilities to support CCDR requirements against a variety of threats to national security. These include invasion, cyberspace attack, and air and missile attacks. Support to HD provided by the National Guard will be IAW DODD 3160.01, *Homeland Defense Activities Conducted by the National Guard.*

(2) **DSCA**

(a) DSCA is support provided by US federal military forces; DOD civilians, DOD contract personnel, DOD component assets, DOD agencies, and National Guard forces (when SecDef, in coordination with the governors of the affected states, elects and requests to use those forces in Title 32, USC status) in response to requests for assistance from civil authorities for domestic emergencies, law enforcement support, and other domestic activities, or from qualifying entities for special events. For DSCA operations, DOD supports and does not supplant civil authorities. The majority of DSCA operations

Chapter VII

are conducted IAW the NRF, which establishes a comprehensive, national, all-hazards approach to domestic incident response. Within a state, that state's governor is the key decision maker and commands the state's National Guard forces when they are not in federal Title 10, USC, status. When the governor mobilizes the National Guard, it will most often be under state active duty when supporting civil authorities.

(b) Other DSCA operations can include CD activities, support to national special security events, or other support to civilian law enforcement IAW specific DOD policies and US law. Commanders and staffs must carefully consider the legal and policy limits imposed on intelligence activities in support of LEAs, and on intelligence activities involving US citizens and entities by intelligence oversight regulations, policies, and executive orders.

(3) **Global Perspective.** Commander, US Northern Command, and Commander, US Pacific Command, have specific responsibilities for HD and DSCA. These responsibilities include conducting operations to deter, prevent, and defeat threats and aggression aimed at the US, its territories, and interests within their assigned AORs, as directed by the President or SecDef. However, DOD support to HD is global in nature and is often conducted by all CCDRs beginning at the source of the threat. In the forward regions outside US territories, the objective is to detect and deter threats to the homeland before they arise and to defeat these threats as early as possible when so directed.

For more information on DSCA, see JP 3-28, Defense Support of Civil Authorities, *and for detailed guidance on DSCA, see DODD 3025.18,* Defense Support of Civil Authorities.

For more information on National Guard support, see DODD 3025.18, Defense Support of Civil Authorities, *and DODI 3025.22,* The Use of the National Guard for Defense Support of Civil Authorities.

For detailed guidance on HD, see JP 3-27, Homeland Defense.

4. Other Considerations

a. **Duration and End State.** Crisis response and limited contingency operations may be relatively short in duration (e.g., NEO, strike, raid) or last for an extended period to achieve the national objective (such as US participation with ten other nations in the independent [non-UN] peacekeeping operation, *Multinational Force and Observers,* in the Sinai Peninsula since 1982). Short duration operations are not always possible, particularly in situations where destabilizing conditions have existed for years or where conditions are such that a long-term commitment is required to achieve national strategic objectives. Nevertheless, it is imperative to have clear national objectives for all types of contingencies.

b. **Intelligence.** As soon as practical, JFCs and their staffs determine intelligence requirements to support the anticipated operation. Intelligence planners also consider the capability for a unit to receive external intelligence support, the capability to store intelligence data, the timeliness of intelligence products, the availability of intelligence publications, and the possibility of using other agencies and organizations as intelligence sources. In some contingencies (e.g., PKO), the term **information collection** is used rather than the term **intelligence** because of the operation's sensitivity.

(1) HUMINT may often provide the most useful source of information and is essential to understanding an enemy or adversary. If a HUMINT infrastructure is not in place when US forces arrive, it needs to be established as quickly as possible. HUMINT also complements other intelligence sources with information not available through technical means. For example, while overhead imagery may graphically depict the number of people gathered in a town square, it cannot gauge the motivations or enthusiasm of the crowd. Additionally, in underdeveloped areas, belligerent forces may not rely heavily on radio communications, thereby denying US forces intelligence derived through signal intercept.

(2) Where there is little USG or US military presence, open-source intelligence (OSINT) may be the best immediately available information to prepare US forces to operate in a foreign country. OSINT from broadcasts, print media, and social networks may be the best immediately available information to provide tip-offs for HUMINT and other intelligence and information collection methods.

(3) Tailored products based on continuous JIPOE can promote timely and comprehensive understanding of all aspects of the OE needed for crisis response and limited contingency operations.

(4) GEOINT consists of the exploitation and analysis of imagery and geospatial information to describe, assess, and visually depict physical features and geographically referenced activities. GEOINT consists of imagery, imagery intelligence, and geospatial information.

For further guidance on GEOINT, refer to JP 2-03, Geospatial Intelligence in Joint Operations.

(5) Intelligence organizations (principally at the JTF HQ level) should include **foreign area officers**. Due to extensive training and experience working in foreign countries as defense attachés and in defense support to US embassy operations, foreign area officers add valuable cultural awareness and insights to intelligence products.

For further guidance on JIPOE, refer to JP 2-01.3, Joint Intelligence Preparation of the Operational Environment.

c. **Operational Limitations.** A JFC tasked with conducting or supporting a crisis response or limited contingency operation may face numerous constraints, restraints, and ROE based on the specific circumstances. For example, international acceptance of each operation may be extremely important, not only because military forces may be used to support international sanctions, but also because of the probability of involvement by international organizations. As a consequence, legal and fiscal constraints unique to the operation should be addressed in detail by the CCDR's staff. Also, operational limitations imposed on any agency or organization involved in the operation should be clarified for other agencies and organizations to facilitate coordination.

d. **Force Protection.** Limited contingency operations may involve a requirement to protect nonmilitary personnel. In the absence of the rule of law, the JFC must address when, how, and to what extent he will extend force protection to civilians and what that protection means.

e. **Training.** Participation in certain types of smaller-scale contingencies may preclude normal mission-related training. For example, infantry units or fighter squadrons conducting certain protracted PO may not have the time, facilities, or environment in which to maintain individual or unit proficiency for traditional missions. In these situations, commanders should develop programs that enable their forces to maintain proficiency in their core competencies/mission essential tasks to the greatest extent possible.

CHAPTER VIII
LARGE-SCALE COMBAT OPERATIONS

> *"Your task will not be an easy one. Your enemy is well trained, well equipped and battle hardened. He will fight savagely.*
>
> *But this is the year 1944! Much has happened since the Nazi triumphs of 1940-41. The United Nations have inflicted upon the Germans great defeats, in open battle, man-to-man. Our air offensive has seriously reduced their strength in the air and their capacity to wage war on the ground. Our Home Fronts have given us an overwhelming superiority in weapons and munitions of war, and placed at our disposal great reserves of trained fighting men. The tide has turned! The free men of the world are marching together to Victory!*
>
> *I have full confidence in your courage and devotion to duty and skill in battle. We will accept nothing less than full Victory!"*
>
> **General Dwight D. Eisenhower D-Day Message**
> **Order of the Day: 6 June 1944**

1. Introduction

a. Traditionally, campaigns are the most extensive joint operations, in terms of the amount of forces and other capabilities committed and duration of operations. In the context of large-scale combat, a campaign is a series of related major operations aimed at achieving strategic and operational objectives within a given time and space. A major operation is a series of tactical actions, such as battles, engagements, and strikes, and is the primary building block of a campaign. Major operations and campaigns typically include multiple phases (e.g., the 1990-1991 Operations DESERT SHIELD and DESERT STORM and 2003 OIF). Campaign planning is appropriate when the contemplated military operations exceed the scope of a single major operation.

b. Campaigns can occur across the continuum of conflict. In campaigns characterized by combat, the general goal is to prevail against the enemy as quickly as possible; conclude hostilities; and establish conditions favorable to the HN, the US, and its multinational partners. Establishing these conditions may require joint forces to conduct stability activities to restore security, provide essential services and humanitarian relief, and conduct emergency reconstruction. **Some crisis-response or contingency operations may not involve large-scale combat, but could meet the definition of a major operation or campaign based on their scale and duration** (e.g., the Tsunami relief efforts in Indonesia or Hurricane Katrina relief efforts in the US, both in 2005).

c. **Campaigns are joint in nature—functional and Service components of the joint force conduct supporting operations, not independent campaigns.** Within a campaign, forces of a single or several Services, coordinated in time and space, conduct operations to achieve strategic or operational objectives in one or more OAs. Forces operate simultaneously or sequentially IAW a common plan, and are controlled by a single Service commander or the JFC.

Chapter VIII

2. Combatant Command Planning

a. The CCMD strategy links national strategic guidance to development of CCMD campaign and contingency plans. A CCMD strategy is a broad statement of the GCC's long-term vision for the AOR and the FCC's long-term vision for the global employment of functional capabilities. CCDRs prepare these strategies in the context of SecDef's priorities outlined in the GEF and the CJCS's objectives articulated in the NMS. However, the size, complexity, and anticipated duration of operations typically magnify the planning challenges. There are three categories of campaigns, which differ generally in scope and focus.

b. CCDRs document the full scope of their campaigns in the set of plans that includes the theater or functional campaign plan, and all of its GEF- and JSCP-directed plans, subordinate and supporting plans, posture or master plans, country plans (for the geographic CCMDs), OPLANs of operations currently in execution, contingency plans, and crisis action plans.

(1) GCCs, as directed in the UCP, GEF, and JSCP, prepare TCPs in APEX OPLAN format for their UCP-assigned AOR and integrate the planning of designated missions assigned to specified CCDRs into their TCPs.

(2) FCCs, as directed in the UCP, GEF, and JSCP, prepare functional campaign plans in APEX OPLAN format for their UCP-assigned missions and responsibilities and synchronize planning across CCMDs, Services, and DOD agencies for designated missions.

(3) The scale and projected duration of a subordinate JFC's crisis response or contingency mission may require the GCC or JFC to develop a campaign plan subordinate to the GCC's theater campaign.

For detailed guidance on joint planning and campaign plans, see JP 5-0, Joint Planning.

THE GULF WAR, 1990-1991

On 2 August 1990, Iraq invaded and occupied Kuwait. Much of the rest of the world, including most other Arab nations, united in condemnation of that action. On 7 August, the operation known as DESERT SHIELD began. Its principal objectives were to deter further aggression and to force Iraq to withdraw from Kuwait. The United Nations (UN) Security Council passed a series of resolutions calling for Iraq to leave Kuwait, finally authorizing "all necessary means," including the use of force, to force Iraq to comply with UN resolutions.

The US led in establishing a political and military coalition to force Iraq from Kuwait and restore stability to the region. The military campaign to accomplish these ends took the form of a series of major operations. These

operations employed the entire capability of the international military coalition and included operations in war and operations other than war throughout.

The campaign—which included Operations DESERT SHIELD and DESERT STORM and the subsequent period of post-conflict operations—can be viewed in the following major phases.

- DEPLOYMENT AND FORCE BUILDUP (to include crisis action planning, mobilization, deployment, and deterrence)

- DEFENSE (with deployment and force buildup continuing)

- OFFENSE

- POSTWAR OPERATIONS (to include redeployment)

DEPLOYMENT AND FORCE BUILDUP. While diplomats attempted to resolve the crisis without combat, the coalition's military forces conducted rapid planning, mobilization, and the largest strategic deployment since World War II. One of the earliest military actions was a maritime interdiction of the shipping of items of military potential to Iraq.

The initial entry of air and land forces into the theater was unopposed. The Commander, United States Central Command (CDRUSCENTCOM), balanced the arrival of these forces to provide an early, viable deterrent capability and the logistic capability needed to receive, further deploy, and sustain the rapidly growing force. Planning, mobilization, and deployment continued throughout this phase.

DEFENSE. While even the earliest arriving forces were in a defensive posture, a viable defense was possible only after the buildup of sufficient coalition air, land, and maritime combat capability. Mobilization and deployment of forces continued. Operations security (OPSEC) measures, operational military deception, and operational psychological operations were used to influence Iraqi dispositions, expectations, and combat effectiveness and thus degrade their abilities to resist CDRUSCENTCOM's selected course of action before engaging enemy forces. This phase ended on 17 January 1991, when Operation DESERT STORM began.

OFFENSE. Operation DESERT STORM began with a major airpower effort—from both land and sea—against strategic targets; Iraqi air, land, and naval forces; logistic infrastructure; and command and control (C2). Land and special operations forces supported this air effort by attacking or designating for attack forward-based Iraqi air defense and radar capability. The objectives of this phase were to gain air supremacy, significantly degrade Iraqi C2, deny information to enemy commanders, destroy enemy forces and infrastructure, and deny freedom of movement. This successful air operation would establish the conditions for the attack by coalition land forces.

Chapter VIII

> While airpower attacked Iraqi forces throughout their depth, land forces repositioned from deceptive locations to attack positions using extensive OPSEC measures and simulations to deny knowledge of movements to the enemy. Coalition land forces moved a great distance in an extremely short time to positions from which they could attack the more vulnerable western flanks of Iraqi forces. US amphibious forces threatened to attack from eastern seaward approaches, drawing Iraqi attention and defensive effort in that direction.
>
> On 24 February, land forces attacked Iraq and rapidly closed on Iraqi flanks. Under a massive and continuous air operation, coalition land forces closed with the Republican Guard. Iraqis surrendered in large numbers. To the extent that it could, the Iraqi military retreated. Within 100 hours of the start of the land force attack, the coalition achieved its strategic objectives and a cease-fire was ordered.
>
> **POSTWAR OPERATIONS.** Coalition forces consolidated their gains and enforced conditions of the cease-fire. The coalition sought to prevent the Iraqi military from taking retribution against its own dissident populace. Task Force Freedom began operations to rebuild Kuwait City.
>
> The end of combat operations did not bring an end to conflict. The coalition conducted peace enforcement operations, humanitarian relief, security operations, extensive weapons and ordnance disposal, and humanitarian assistance. On 5 April, for example, President Bush announced the beginning of a relief operation in the area of northern Iraq. By 7 April, US aircraft from Europe were dropping relief supplies over the Iraqi border. Several thousand Service personnel who had participated in Operation DESERT STORM eventually redeployed to Turkey and northern Iraq in this joint and multinational relief operation.
>
> This postwar phase also included the major operations associated with the redeployment and demobilization of forces.
>
> **Various Sources**

3. **Setting Conditions for Theater Operations**

   a. **General.** CCDRs and JFCs execute their campaigns and operations in pursuit of US national objectives and to shape the OE. In pursuit of national objectives, these campaigns and operations also seek to prevent, prepare for, or mitigate the impact of a crisis or contingency. In many cases, these actions enhance bonds between potential multinational partners, increase understanding of the region, help ensure access when required, and strengthen the capability for future multinational operations, all of which help prevent crises from developing.

   b. **Organizing and Training Forces.** Organizing and, where possible, training forces to conduct operations throughout the OA can be a deterrent. JTFs and components that are likely to be employed in theater operations should be exercised regularly during peacetime.

Staffs should be identified and trained for planning and controlling joint and multinational operations. The composition of joint force staffs should reflect the composition of the joint force to ensure those employing joint forces have thorough knowledge of their capabilities and limitations. When possible, JFCs and their staffs should invite non-DOD agencies to participate in training to facilitate a common understanding and to build a working relationship prior to actual execution. Commanders must continue to refine interactions with interagency partners they will work with most often and develop common procedures to improve interoperability. When it is not possible to train forces in the theater of employment, as with US-based forces with multiple tasks, commanders should make maximum use of regularly scheduled and ad hoc exercise opportunities. The training focus for all forces and the basis for exercise objectives should be the CCDR's joint mission-essential tasks.

c. **Rehearsals.** Rehearsal provides an opportunity to learn, understand, and practice a plan in the time available before actual execution. Rehearsing key combat and sustainment actions allows participants to become familiar with the operation, visualize the plan, and identify possible friction points. This process orients joint and multinational forces to surroundings and to other units during execution. Rehearsals also provide a forum for subordinate leaders to analyze the plan, but they must exercise caution in adjusting the plan. Changes must be coordinated throughout the chain of command to prevent errors in integration and synchronization. HQ at the tactical level often conduct rehearsals involving participation of maneuver forces positioned on terrain that mirrors the OE. HQ at the operational level rehearse key aspects of a plan using command post exercises, typically supported by computer-aided simulations. While the joint force may not be able to rehearse an entire operation, the JFC should identify essential elements for rehearsal.

d. **Maintaining Operational Access.** JFCs must overcome the enemy's A2/AD capabilities to establish and maintain access to OAs where they are likely to operate, ensuring forward presence, basing (to include availability of airfields and seaports and adequate sustainment), resiliency of combat power after enemy action, freedom of navigation, and cooperation with allied and/or coalition nations to enhance operational reach. In part, this effort is national or multinational, involving maintenance of intertheater (between theaters) air, land, sea, space, EMS, and cyberspace LOCs. Supporting CCDRs can greatly enhance this effort.

e. **Space Considerations.** Space operations support all joint operations. When conflict occurs, commanders need to ensure US, allied, and/or multinational forces gain and maintain space superiority, which is achieved through space situational awareness, space force enhancement, space support, space control, and space force application. Commanders must anticipate hostile actions that may affect friendly space operations. Commanders should also anticipate the proliferation and increasing sophistication of commercial space capabilities and products available that the commander can leverage, but which may also be available to enemies and adversaries. USSTRATCOM plans and conducts space operations. **The GCC, in coordination with other USG departments and agencies, conducts certain aspects of theater space operations to include planning for, supporting, and conducting the recovery of astronauts, space vehicles, space payloads, and objects as directed.** They may also request the CDRUSSTRATCOM's

Chapter VIII

assistance in integrating space forces, capabilities, and considerations into each phase of campaign and major OPLANs. Global and theater space operations require robust planning and skilled employment to synchronize and integrate space operations with the joint operation. It is therefore incumbent upon the GCCs to coordinate as required to minimize conflicts. Space capabilities help shape the OE in a variety of ways including providing intelligence and communications necessary to keep commanders and leaders informed worldwide. JFCs and their components should request space support early in the planning process to ensure effective and efficient use of space assets.

For further guidance on space operations, refer to JP 3-14, Space Operations, *and Air Force Doctrine Annex 3-14,* Space Operations.

f. **EMS Considerations**. The joint force is critically dependent on the EMS for operations across all joint functions and throughout the OE. For example, modern C2 requires operation of EMS-dependent sensing and communication systems, while advanced weapons rely on positioning, navigation, and timing (PNT) information transmitted through the EMS. Therefore, the joint force must obtain EMS superiority as an essential prerequisite to executing joint operations. EMS superiority is that degree of dominance in the EMS that permits the conduct of operations at a given time and place without prohibitive interference, while affecting an adversary's ability to do the same. Achieving EMS superiority is complicated by increasing joint EMS-use requirements, EME congestion, and proliferation of EMS threats. Joint forces execute JEMSO, facilitated by electromagnetic battle management (EMBM), in order to achieve the necessary unity of effort for EMS superiority.

For further information on EMS/JEMSO, see the National Military Strategic Plan for Electronic Warfare; *JP 6-01,* Joint Electromagnetic Spectrum Management Operations; *JP 3-13.1,* Electronic Warfare; *and the* DOD EMS Strategy.

g. **Stability Activities**. Activities in the shape phase may focus on continued planning and preparation for anticipated stability activities in the subsequent phases. These activities should include conducting collaborative interagency planning to synchronize the civil-military effort, confirming the feasibility of pertinent military objectives and the military end state, and providing for adequate intelligence, an appropriate force mix, and other capabilities. US military support to stabilization efforts in this phase may be required as part of the USG's security sector assistance, purposed to quickly restore security and infrastructure or provide humanitarian relief in select portions of the OA to dissuade further adversary actions or to help gain and maintain access and future success.

4. Considerations for Deterrence

a. **General.** The deter phase is characterized by preparatory actions that indicate resolve to commit resources and respond to the situation. These actions begin when a CCDR or JFC identifies that routine operations may not achieve desired objectives due to an adversary's actions. This requires the commander to have identified CCIRs and assessed whether additional resources, outside those currently allocated and assigned for ongoing operations, are required to defuse the crisis, reassure partners, demonstrate the intent to deny the adversary's goals, and execute subsequent phases of the operation. Deterrence should be based on capability (having the means to influence behavior), credibility (maintaining a level of believability that the proposed actions may actually be employed), and communication (transmitting the intended message to the desired audience) to ensure greater effectiveness (effectiveness of deterrence must be viewed from the perspective of the agent/actor that is to be deterred). Before hostilities begin, the JFC and staff analyze and assess the adversary's goals and decision-making process to determine how, where, and when these can be affected and what friendly actions (military and others) can influence events and act as a deterrent. For example, traditional US military force may be less of a deterrent to irregular and non-state threats, and the assessment should therefore consider alternative or complementary approaches. Emphasis should be placed on setting the conditions for successful joint operations in the dominate and follow-on phases should deterrence fail.

b. **Preparing the OA**

(1) **Special Operations.** SOF play a major role in preparing and shaping the operational area and environment by setting conditions which mitigate risk and facilitate successful follow-on operations. The regional focus, cross-cultural/ethnic insights, language capabilities, and relationships of SOF provide access to and influence in nations where the presence of conventional US forces is unacceptable or inappropriate. SOF contributions can provide operational leverage by gathering critical information, undermining an adversary's will or capacity to wage war, and enhancing the capabilities of conventional US, multinational, or indigenous/surrogate forces. CDRUSSOCOM synchronizes planning for global operations against terrorist networks in coordination with other CCMDs, the Services, and, as directed, appropriate USG departments and agencies. In coordination with GCCs and the supported JFC, CDRUSSOCOM plans and executes or synchronizes the execution of activities related to preparing the OE and OA, or provides SOF to other CCMDs.

For further guidance on special operations, refer to JP 3-05, Special Operations.

(2) **Stability Activities.** Joint force planning and operations conducted prior to commencement of hostilities should establish a sound foundation for operations in the stabilize and enable civil authority phases. JFCs should anticipate and address how to fill the power vacuum created when sustained combat operations wind down. Accomplishing this task should ease the transition to operations in the stabilize phase and shorten the path to national objectives and transition to another authority. Considerations include actions to:

(a) Limit the damage to key infrastructure (water, energy, medical) and services.

(b) Assist with the restoration and development of power generation facilities.

Chapter VIII

(c) Establish the intended disposition of captured leadership and demobilized military and paramilitary forces.

(d) Provide for the availability of cash or other means of financial exchange.

(e) Determine the proper force mix (e.g., combat, military police, CA, engineer, medical, multinational).

(f) Assess availability of HN law enforcement and health and medical resources.

(g) Secure key infrastructure nodes and facilitate HN law enforcement and first responder services.

(h) Develop and disseminate information necessary to suppress potential new enemies and promote new governmental authority.

(3) **CA** forces have a variety of specialty skills that may support the joint operation being planned. CA forces conduct military engagement, humanitarian and civic assistance, and nation assistance to influence HN and foreign nation populations. CA forces assess impacts of the population and culture on military operations, assess impact of military operations on the population and culture, and facilitate interorganizational coordination. Establishing and maintaining civil-military relations may include interaction among US, allied, multinational, and HN forces, as well as other government agencies, international organizations, and NGOs. CA forces can provide expertise on factors that directly affect military operations to include culture, social structure, economic systems, language, and HNS capabilities. CA may be able to perform functions for limited durations that normally are the responsibility of local or indigenous governments. Employment of CA forces should be based upon a clear concept of CA mission requirements for the type operation being planned.

For further guidance on CA, refer to JP 3-57, Civil-Military Operations.

(4) **Sustainment.** Thorough planning for logistic and personnel support is critical. Planning must include active participation by all deploying and in-theater US and multinational forces, as well as interagency personnel. This planning is done through theater distribution plans (TDPs) in support of the GCCs' TCPs. Setting the conditions enables the JFCs to address global, end-to-end distribution requirements and identify critical capabilities, infrastructure, and relationships required to be resourced and emplaced in a timely manner to sustain and enable global distribution operations. USTRANSCOM synchronizes efforts between global TDPs in support of the Campaign Plan for Global Distribution. The GCC conducts theater distribution operations, and the GCC may request USTRANSCOM's assistance in integrating distribution enablers, capabilities, and considerations into each phase of the campaign and contingency plans. Global and theater distribution operations require robust planning and skilled employment to synchronize and integrate distribution operations with the joint operation.

For further guidance on sustainment and distribution operations, refer to JP 4-0, Joint Logistics, *and JP 4-09,* Distribution Operations.

c. **Isolating the Enemy**

(1) With Presidential and SecDef approval, guidance, and national support, **JFCs strive to isolate enemies by denying them allies and sanctuary.** The intent is to strip away as much enemy support or freedom of action as possible, while limiting the enemy's potential for horizontal or vertical escalation. JFCs may also be tasked by the President and SecDef to support diplomatic, economic, and informational actions.

(2) **The JFC also seeks to isolate the main enemy force from both its strategic leadership and its supporting infrastructure.** Such isolation can be achieved through the use of IRCs and the interdiction of LOCs or resources affecting the enemy's ability to conduct or sustain military operations. This step serves to deny the enemy both physical and psychological support and may separate the enemy leadership and military from their public support.

d. **FDOs and FROs.** FDOs and FROs are executed on order and provide scalable options to respond to a crisis. Both provide the ability to scale up (escalate) or de-escalate based on continuous assessment of an adversary's actions and reaction. While FDOs are primarily intended to prevent the crisis from worsening and allow for de-escalation, FROs are generally punitive in nature.

(1) **FDOs.** FDOs are preplanned, deterrence-oriented actions carefully tailored to bring an issue to early resolution without armed conflict. Both military and nonmilitary FDOs can be used to dissuade actions before a crisis arises or to deter further aggression during a crisis. FDOs are developed for each instrument of national power, but they are most effective when used in combination.

(a) **Military FDOs** can be initiated before or after unambiguous warning of enemy action. Deployment timelines, combined with the requirement for a rapid, early response, generally require economy of force; however, military FDOs should not increase risk to the force that exceeds the potential benefit of the desired effect. Military FDOs must be carefully tailored regarding timing, efficiency, and effectiveness. They can rapidly improve the military balance of power in the OA, especially in terms of early warning, intelligence gathering, logistic infrastructure, air and maritime forces, MISO, and protection without precipitating armed response from the adversary. Care should be taken to avoid undesired effects such as eliciting an armed response should adversary leadership perceive that friendly military FDOs are being used as preparation for a preemptive attack.

(b) **Nonmilitary FDOs** are preplanned, preemptive actions taken by other government agencies to dissuade an adversary from initiating hostilities. Nonmilitary FDOs need to be coordinated, integrated, and synchronized with military FDOs to focus all instruments of national power.

(2) **FROs.** FROs, usually used in response to terrorism, can also be employed in response to aggression by a competitor or adversary. Like FDOs, the discussion should

Chapter VIII

include indicators of their effectiveness and probability of consequences, desired and undesired. The basic purpose of FROs is to preempt and/or respond to attacks against the US and/or US interests. FROs are intended to facilitate early decision making by developing a wide range of prospective actions carefully tailored to produce desired effects, congruent with national security policy objectives. A FRO is the venue in which various military capabilities are made available to the President and SecDef, with actions appropriate and adaptable to existing circumstances, in reaction to any threat or attack.

For further guidance on planning FDOs and FROs, refer to JP 5-0, Joint Planning.

e. **Protection.** JFCs must protect their forces and their freedom of action to accomplish their mission. This dictates that JFCs not only provide force protection, but be aware of and participate as appropriate in the protection of interagency and regional multinational capabilities and activities. JFCs may spend as much time on protection to assure partners to preserve coalition resolve and maintain access as on direct preparation of their forces for combat.

f. **Space Operations.** JFCs depend upon and exploit the advantages of space capabilities. During the deter phase, space forces are limited to already fielded and immediately deployable assets and established priorities for service. As the situation develops, priorities for space force enhancement may change to aid the JFC in assessing the changing OE. Most importantly, the JFC and component commanders need to anticipate "surge" space capabilities needed for future phases due to the long lead times to reprioritize or acquire additional capabilities.

For further guidance on space operations, refer to JP 3-14, Space Operations, *and Air Force Doctrine Annex 3-14,* Space Operations.

g. **GEOINT Support to Operations.** Geospatial products or services—including maps, charts, imagery products, web services, and support data—must be fully coordinated with JFC components, as well as with the Joint Staff, Office of the Secretary of Defense, and the NGA through the JFC's GEOINT cell. Requests for or updates to GEOINT products, including maps or annotated imagery products, should be submitted as early as possible through the JFC's GEOINT cell to the NGA support team at the JFC's HQ. US products should be used whenever possible, since the accuracy, scale, and reliability of foreign maps and charts may vary widely from US products. In any joint or multinational operation, the World Geodetic System-1984 should be the common system. If US products are to be used in a coalition environment or within a multinational HQ, the release of US mapping materials or imagery products requires foreign disclosure/release adjudication.

For further guidance on GEOINT, refer to JP 2-03, Geospatial Intelligence in Joint Operations.

h. **Physical Environment**

(1) **Weather, terrain, and sea conditions** can significantly affect operations and sustainment support of the joint force and should be carefully assessed before and during sustained combat operations. Mobility of the force, integration and synchronization of

operations, and ability to employ precision munitions can be affected by degraded conditions. Climatological and hydrographic planning tools, studies, and forecast products help the JFC determine the most advantageous time and location to conduct operations.

For additional guidance, refer to JP 3-59, Meteorological and Oceanographic Operations.

(2) **Urban areas** possess all of the characteristics of the natural landscape, coupled with man-made construction and the associated infrastructure, resulting in a complicated and dynamic environment that influences the conduct of military operations in many ways. **The most distinguishing characteristic of operations in urban areas, however, is not the infrastructure but the density of civilians.** Joint urban operations (JUOs) are conducted in large, densely populated areas with problems unique to clearing enemy forces, while possibly restoring services and managing major concentrations of people. For example, industrial areas and port facilities often are collocated with highly populated areas, creating the opportunity for accidental or deliberate release of toxic industrial materials which could impact JUOs. During JUOs, joint forces may not focus only on destruction of enemy forces but may also be required to take steps necessary to protect and support civilians and their infrastructure from which they receive services necessary for survival. As such, ROE during JUOs may be more restrictive than for other types of operations. When planning JUOs, the JFC and staff should consider the impact of military operations on civilians to include their culture, values, and infrastructure; thereby viewing the urban area as a dynamic and complex system—not solely as terrain. This implies the joint force should be capable of understanding the specific urban environment; sensing, locating, and isolating the enemy among civilians; and applying combat power precisely and discriminately.

For additional guidance on JUOs, refer to JP 3-06, Joint Urban Operations.

(3) **Littoral Areas.** The littoral area contains two parts. First is the seaward area from the open ocean to the shore, which must be controlled to support operations ashore. Second is the landward area inland from the shore that can be supported and defended directly from the sea. Control of the littoral area often is essential to maritime superiority. Maritime operations conducted in the littoral area can project power, fires, and forces to achieve the JFC's objectives. These operations facilitate the entry and movement of other elements of the joint force through the seizure of an enemy's port, naval base, or air base. Access to, and freedom of maneuver within, the littorals depend on the enemy's A2/AD actions and capabilities, such as the use of surface-to-surface missiles and use of mines. The ability to avoid detection and maneuver to evade can be reduced in the littorals due to the proximity to enemy sensors and the lack of maneuver space, thus increasing risk. Depending on the situation, mine warfare may be critical to control of the littoral areas.

(4) **The EMS,** which has become increasingly complex, contested, and congested as technology has advanced, can significantly affect joint force operations. Operational experiences demonstrate not only how successful control of the EMS can influence the outcome of the conflict, but highlight US vulnerabilities and dependence on the EMS to successfully operate. When planning operations, the JFC should consider both the impact of military operations on the current EME, as well as its effect on military operations. Key

Chapter VIII

to this is the JFC's execution of JEMSO, facilitated by EMBM, in order to achieve integration. This will ensure EMS-dependent systems are mission ready and compatible within the intended EME.

5. Considerations for Seizing the Initiative

a. **General.** As operations commence, the JFC needs to exploit friendly advantages and capabilities to shock, demoralize, and disrupt the enemy immediately. The JFC seeks decisive advantage through the use of all available elements of combat power to seize and maintain the initiative, deny the enemy the opportunity to achieve its objectives, and generate in the enemy a sense of inevitable failure and defeat. Additionally, the JFC coordinates with other USG departments and agencies to facilitate coherent use of all instruments of national power in achieving national strategic objectives.

b. **Force Projection**

(1) Projecting US military force invariably requires extensive use of the international waters, international airspace, space, cyberspace, and the EMS to gain operational access. Our ability to freely maneuver to position and sustain our forces is vital to our national interests and those of our PNs. US forces may gain operational access to areas through invitation by an HN to establish an operating base in or near the conflict or by the use of forcible entry operations. Treaties, agreements, and activities that occur during the shape and deter phases may aid in the invitation to establish a base or support facility. However, gaining and maintaining operational access requires the ability to defeat the enemy's A2/AD actions and capabilities.

(2) The President and SecDef may direct a CCDR to resolve a crisis quickly, employing immediately available forces and appropriate FDOs as discussed above to preclude escalation. When these forces and actions are not sufficient, follow-on strikes and/or the **deployment** of forces from CONUS or another theater and/or the use of multinational forces may be necessary. Consequently, the CCDR must sequence, enable, and protect the deployment of forces to create early decisive advantage. The CCDR should not overlook enemy A2/AD capabilities that may affect the deployment of combat and logistic forces from bases to ports of embarkation. The CCDR may have to adjust the time-phased force and deployment data to meet a changing OE. The deployment of forces may be either opposed or unopposed by an enemy.

(a) **Opposed.** Initial operations may be designed to suppress enemy A2/AD capabilities. For example, the ability to generate sufficient combat power through long-range air operations or from the sea can provide for effective force projection in the absence of timely or unencumbered access. Other opposed situations may require a forcible entry capability. In other cases, force projection can be accomplished rapidly by forcible entry operations coordinated with strategic air mobility, sealift, and pre-positioned forces. For example, the seizure and defense of lodgment areas by amphibious forces would then serve as initial entry points for the continuous and uninterrupted flow of forces and materiel into the theater. Both efforts demand a versatile mix of forces that are organized, trained, equipped, and poised to respond quickly.

(b) **Unopposed** deployment operations provide the JFC and subordinate components a more flexible OE to efficiently and effectively build combat power, train, rehearse, acclimate, and otherwise establish the conditions for successful combat operations. In unopposed entry, JFCs arrange the flow of forces, to include significant theater opening logistics forces, that best facilitates the CONOPS. In these situations, logistics forces may be a higher priority for early deployment than combat forces, as determined by the in-theater protection requirements.

(3) Commanders should brief deploying forces on the threat and force protection requirements prior to deployment and upon arrival in the OA. Also, JFCs and their subordinate commanders evaluate the timing, location, and other factors of force deployment in each COA for the impact of sabotage, criminal activity, and terrorist acts and their impact on joint reception, staging, onward movement, and integration (JRSOI) and the follow-on CONOPS. The threat could involve those not directly supporting or sympathetic to the enemy, but those seeking to take advantage of the situation. When the situation involves a failed or failing WMD-possessor state, commanders should consider that opportunists could employ lost or stolen WMD.

(4) During force projection, US forces and ports of debarkation must be protected. JFCs should introduce forces in a manner that provides security for rapid force buildup. Therefore, early entry forces should deploy with sufficient organic and supporting capabilities to preserve their freedom of action and protect personnel and equipment from potential or likely threats. Early entry forces also should include a deployable joint C2 capability to rapidly assess the situation, make decisions, and conduct initial operations.

(5) JRSOI occurs in the OA and comprises the essential processes required to transition arriving personnel, equipment, and materiel into forces capable of meeting operational requirements. Forces are vulnerable during JRSOI, so planning must include force protection requirements.

For further information on JRSOI, refer to JP 3-35, Deployment and Redeployment Operations.

c. **Unit Integrity During Deployment**

(1) US military forces normally train as units and are best able to accomplish a mission when deployed intact. By deploying as an existing unit, forces are able to continue to operate under established procedures, adapting them to the mission and situation, as required. When personnel and elements are drawn from various commands, effectiveness may be decreased. By deploying without established operating procedures, an ad hoc force takes more time to form and adjust to requirements of the mission. This not only complicates mission accomplishment, but also may have an impact on force protection.

(2) Even if diplomatic/political restraints on an operation dictate that a large force cannot be deployed intact, commanders should select elements for deployment that have established internal procedures and structures, have trained and operated together, and possess appropriate joint force combat capabilities. In order to provide a JFC with needed

Chapter VIII

versatility, it may not be possible to preserve complete unit integrity. In such cases, units must be prepared to send elements that are able to operate independently of parent units. Attachment to a related unit is the usual mode. In this instance, units not accustomed to having attachments may be required to provide administrative and logistic support to normally unrelated units.

(3) The CCDR, in coordination with Commander, USTRANSCOM; subordinate JFCs; and the Service component commanders, needs to carefully balance the desire to retain unit integrity through the deployment process with the effective use of strategic lift platforms. While maximizing unit integrity may reduce JRSOI requirements and allow combat units to be employed more quickly, doing so will often have a direct negative impact on the efficient use of the limited strategic lift. In some cases, this negative impact on strategic lift may have a negative effect on DOD deployment and sustainment requirements beyond the GCC's AOR. A general rule of thumb is that unit integrity is much more important for early deploying units than for follow-on forces.

d. While access operations focus on enabling access to the OA, **entry operations focus on actions within the OA.** Joint forces conduct entry operations for various purposes, including to defeat threats to the access and use of portions of the OE; to control, defeat, disable, and/or dispose of specific WMD threats; to assist populations and groups; to establish a lodgment; and to conduct other limited duration missions.

e. **Entry operations may be unopposed or opposed.** Unopposed entry operations often, but not always, follow unopposed access. These circumstances generally allow orderly deployment into the OA in preparation for follow-on operations. **Forcible entry** is a joint military operation conducted either as a major operation or a part of a larger campaign to **seize and hold a military lodgment in the face of armed opposition** for the continuous landing of forces. Forcible entry operations can strike directly at the enemy COGs and can open new avenues for other military operations.

(1) Forcible entry operations may include amphibious, airborne, and air assault operations, or any combination thereof. Forcible entry operations can create multiple dilemmas by creating threats that exceed the enemy's capability to respond. Commanders will employ distributed, yet coherent, operations to attack the objective area or areas. The net result will be a coordinated attack that overwhelms the enemy before they have time to react. A well-positioned and networked force enables the defeat of any enemy reaction and facilitates follow-on operations, if required.

(2) **Forcible entry is normally complex and risky** and should, therefore, be kept as simple as possible in concept. These operations require extensive intelligence, detailed coordination, innovation, and flexibility. Schemes of maneuver and coordination between forces need to be clearly understood by all participants. Forces are tailored for the mission and echeloned to permit simultaneous deployment and employment. When airborne, amphibious, and air assault operations are combined, unity of command is vital. Rehearsals are a critical part of preparation for forcible entry. Participating forces need to be prepared to fight immediately upon arrival and require robust communications and intelligence capabilities to move with forward elements.

(3) **The forcible entry force must be prepared to immediately transition to follow-on operations and should plan accordingly.** Joint forcible entry actions occur in both singular and multiple operations. These actions include establishing forward presence, preparing the OA, opening entry points, establishing and sustaining access, receiving follow-on forces, conducting follow-on operations, sustaining the operations, and conducting decisive operations.

(4) **Successful OPSEC and MILDEC may confuse the enemy and ease forcible entry operations.** OPSEC helps foster a credible MILDEC. Additionally, the actions, themes, and messages portrayed by all friendly forces must be consistent if MILDEC is to be believable.

OPERATION JUST CAUSE

In the early morning hours of 20 December 1989, the Commander, US Southern Command, Joint Task Force (JTF) Panama, conducted multiple, simultaneous forcible entry operations to begin Operation JUST CAUSE. By parachute assault, forces seized key lodgments at Torrijos-Tucumen Military Airfield and International Airport and at the Panamanian Defense Forces (PDF) base at Rio Hato. The JTF used these lodgments for force buildup and to launch immediate assaults against the PDF.

The JTF commander synchronized the forcible entry operations with numerous other operations involving virtually all capabilities of the joint force. The parachute assault forces strategically deployed at staggered times from bases in the continental US, some in C-141 Starlifters, others in slower C-130 transport planes. One large formation experienced delays from a sudden ice storm at the departure airfield—its operations and timing were revised in the air. H-hour was even adjusted for assault operations because of intelligence that indicated a possible compromise. Special operations forces (SOF) reconnaissance and direct action teams provided last-minute information on widely dispersed targets.

At H-hour the parachute assault forces, forward-deployed forces, SOF, and air elements of the joint force simultaneously attacked 27 targets—most of them in the vicinity of the Panama Canal. Illustrating that joint force commanders organize and apply force in a manner that fits the situation, the JTF commander employed land and SOF to attack strategic targets and stealth aircraft to attack tactical and operational-level targets.

The forcible entry operations, combined with simultaneous and follow-on attack against enemy command and control facilities and key units, seized the initiative and paralyzed enemy decision making. Most fighting was concluded within 24 hours. Casualties were minimized. It was a classic *coup de main*.

Various Sources

Chapter VIII

(5) **SOF may precede forcible entry forces** to identify, clarify, establish, or modify conditions in the lodgment. SOF may conduct the assaults to seize small, initial lodgments such as airfields or seaports. They may provide or assist in employing fire support and conduct other operations in support of the forcible entry, such as seizing airfields or conducting reconnaissance of landing zones or amphibious landing sites. They may conduct special reconnaissance and direct action well beyond the lodgment to identify, interdict, and destroy forces that threaten the conventional entry force.

(6) **The sustainment requirements and challenges** for forcible entry operations **can be formidable,** but must not be allowed to become such an overriding concern that the forcible entry operation itself is jeopardized. JFCs must carefully balance the introduction of sustainment forces needed to support initial combat with combat forces required to establish, maintain, and protect the lodgment as well as forces required to transition to follow-on operations.

For additional and detailed guidance on forcible entry operations, refer to JP 3-18, Joint Forcible Entry Operations.

f. **Attack of Enemy COGs.** As part of creating decisive advantages early, joint force operations may be directed immediately against enemy COGs using conventional forces and SOF if COGs are vulnerable and sufficient friendly force capabilities are available. These attacks may be decisive or may begin offensive operations throughout the enemy's depth that can create dilemmas causing paralysis and destroying cohesion.

g. **Full-Spectrum Superiority.** Mission success in large-scale combat requires full-spectrum superiority; the cumulative effect of achieving superiority in the air, land, maritime, and space domains; the information environment; and the EMS. Such superiority permits the conduct of joint operations without effective opposition or prohibitive interference. JFCs seek superiority throughout the OE to accomplish the mission as rapidly as possible. The JFC may have to initially focus all available joint forces on seizing the initiative. A delay at the outset of combat may damage US credibility, lessen coalition support, and provide incentives for other adversaries to begin conflicts elsewhere.

(1) **JFCs normally strive to achieve air and maritime superiority early.** Air and maritime superiority allows joint forces to conduct operations without prohibitive interference from opposing air and maritime forces. Control of the air is a critical enabler because it allows joint forces both freedom from attack and freedom to attack. Using both defensive and offensive operations, JFCs employ complementary weapon systems and sensors to achieve air and maritime superiority.

(2) **Land forces** can be moved quickly into an area to deter the enemy from inserting forces, thereby precluding the enemy from gaining an operational advantage. The rapid deployment and employment of land forces (with support of other components) enable sustained operations, more quickly contribute to the enemy's defeat, and help restore stability in the OA.

(3) **Space superiority must be achieved early to support freedom of action.** Space superiority allows the JFC access to communications, environmental monitoring, PNT warning, and intelligence collection assets without prohibitive interference by the opposing force. Space control operations are conducted by joint and allied and/or coalition forces to gain and maintain space superiority.

(4) Early **superiority in the information environment** (which includes cyberspace) is vital in joint operations. **It degrades the enemy's C2 while allowing the JFC to maximize friendly C2 capabilities.** Information superiority also allows the JFC to better understand the enemy's intentions, capabilities, and actions, as well as influence foreign attitudes and perceptions of the operation.

(5) Control of the EME must be achieved early to support freedom of action. This control is important for superiority across the physical domains and information environment.

h. **C2 in Littoral Areas**

(1) **Controlled littoral areas often offer the best positions from which to begin, sustain, and support joint operations,** especially in OAs with limited or poor infrastructure for supporting US joint operations ashore. JFCs can gain and maintain the initiative through the ability to project fires and employ forces from sea-based assets in combination with C2, intelligence collection, and IRCs. Maritime forces operating in littoral areas can dominate coastal areas and rapidly generate high intensity offensive power at times and in locations required by JFCs. Maritime forces' relative freedom of action enables JFCs to position these capabilities where they can readily strike opponents. Maritime forces' very presence, if made known, can pose a threat that the enemy cannot ignore.

(2) **JFCs can operate from a HQ platform at sea.** Depending on the nature of the joint operation, a maritime commander can serve as the JFC or function as a JFACC while the operation is primarily maritime and shift that command ashore if the operation shifts landward IAW the JFC's CONOPS. A sea base provides JFCs with the ability to command and control forces and conduct select functions and tasks at sea without dependence on infrastructure ashore. In other cases, a maritime HQ may serve as the base of the joint force HQ, or subordinate JFCs or other component commanders may use the C2 and intelligence facilities aboard ship.

(3) **Transferring C2 from sea to shore** requires detailed planning, active liaison, and coordination throughout the joint force. Such a transition may involve a simple movement of flags and supporting personnel, or it may require a complete change of joint force HQ. The new joint force HQ may use personnel and equipment, especially communications equipment, from the old HQ, or it may require augmentation from different sources. One technique is to transfer C2 in several stages. Another technique is for the JFC to satellite off the capabilities of one of the components ashore until the new HQ is fully prepared. Whichever way the transition is done, staffs should develop detailed

Chapter VIII

> **SPECIAL OPERATIONS FORCES AND CONVENTIONAL FORCES INTEGRATION DURING OPERATION ENDURING FREEDOM**
>
> **Special operations forces (SOF) and conventional forces integration demonstrated powerful air-ground synergies in Operation ENDURING FREEDOM. While performing the classic core activity of unconventional warfare, Army special forces organized and coordinated operations of the Northern Alliance against the Taliban and al Qaeda. Supported by other joint SOF, they frequently directed massive and effective close air support from Air Force, Navy, and Marine Corps assets. The effects of the continuous SOF-directed air strikes so weakened the Taliban and al Qaeda that the Northern Alliance was able to quickly capture the major cities of Afghanistan early in the campaign.**
>
> **Various Sources**

checklists to address all of the C2 requirements and the timing of transfer of each. The value of joint training and rehearsals in this transition is evident.

i. **SOF-Conventional Force Integration.** The JFC, using SOF independently or integrated with conventional forces, gains an additional and specialized capability to achieve objectives that might not otherwise be attainable. Integration enables the JFC to take fullest advantage of conventional and SOF core competencies. SOF are most effective when special operations are fully integrated into the overall plan and the execution of special operations is through proper SOF C2 elements in a supporting or supported relationship with conventional forces. Joint SOF C2 elements are provided to conduct a specific special operation or prosecute special operations in support of a joint campaign or operation. Special operations commanders also provide liaison to component commands to integrate, coordinate, and deconflict SOF and conventional force operations. Exchange of SOF and conventional force LNOs is essential to enhance situational awareness and reduce risk of friendly fire incidents.

j. **Stability Activities.** Combat in this phase provides an opportunity to begin various stability activities that will help achieve military strategic and operational-level objectives and create the conditions for the later stability and enable civil authority phases. Operations to neutralize or eliminate potential stabilize phase enemies may be initiated. National and local HN authorities may be contacted and offered support. Key infrastructure may be seized or otherwise protected. Civil IM, which is broadly tasked to support the overall intelligence collection on the status of enemy infrastructure, government organizations, and humanitarian needs, should be increased. MISO used to influence the behavior of approved foreign target audiences in support of military strategic and operational objectives can ease the situation encountered when sustained combat is concluded. In coordination with interorganizational participants, the JFC must arrange for necessary financial support of these operations well in advance.

k. **Protection.** JFCs must strive to conserve the fighting potential of the joint/multinational force at the onset of combat operations. Further, HN infrastructure and logistic support key to force projection and sustainment of the force must be protected.

JFCs counter the enemy's fires and maneuver by making personnel, systems, and units difficult to locate, strike, and destroy. They protect their force from enemy maneuver and fires by using various physical and informational measures. OPSEC and MILDEC are key elements of this effort. Operations to gain air, space, maritime, and EMS superiority; defensive use of IO; PR; and protection of airports and seaports, LOCs, and friendly force lodgment also contribute significantly to force protection at the onset of combat operations.

l. **Prevention of Friendly Fire Incidents.** JFCs must make every effort to reduce the potential for the killing or wounding of friendly personnel by friendly fire. The destructive power and range of modern weapons, coupled with the high intensity and rapid tempo of modern combat, increase the potential for friendly fire incidents. Commanders must be aware of those situations that increase the risk of friendly fire incidents and institute appropriate preventive measures. The primary mechanisms for reducing friendly fire incidents are command emphasis, disciplined operations, close coordination among component commands and multinational partners, SOPs, training and exercises, technology solutions (e.g., identify friend or foe, blue force tracking), rehearsals, effective CID, and enhanced awareness of the OE. Commanders should seek to minimize friendly fire incidents while not limiting boldness and initiative. CCMDs should consult with USAID when it has a mission presence to determine locations of friendly international organizations, NGOs, and local partners operating in the targeted area to avoid friendly fire incidents.

6. **Considerations for Dominance**

   a. **General.** JFCs conduct sustained combat operations when a swift victory is not possible. During sustained combat operations, JFCs simultaneously employ conventional forces and SOF throughout the OA. The JFC may designate one component or LOO to be the main effort, with other components providing support and other LOOs as supporting efforts. When conditions or plans change, the main effort might shift. Some missions and operations (i.e., strategic attack, interdiction, and IO) continue throughout to deny the enemy sanctuary, freedom of action, or informational advantage. These missions and operations, when executed concurrently with other operations, degrade enemy morale and physical cohesion and bring the enemy closer to culmination. When prevented from concentrating, opponents can be attacked, isolated at tactical and operational levels, and defeated in detail. At other times, JFCs may cause their opponents to concentrate their forces, facilitating their attack by friendly forces. In some circumstances (e.g., regime change, ensuring stability prior to transition to civil authority), the JFC may be required to maintain a temporary military occupation of enemy territory while continuing offensive actions. If the occupation is extended and a country's government is not functioning, the JFC may be required to establish a military government through the designation of a transitional military authority.

   b. **Operating in the Littoral Areas.** Even when joint forces are firmly established ashore, littoral operations provide JFCs with excellent opportunities to gain leverage over the enemy by operational maneuver from the sea. Such operations can introduce significant size forces over relatively great distances in short periods of time into the rear or flanks of the enemy. The mobility and fire support capability of maritime forces at sea, coupled with the ability to rapidly land operationally significant forces, can be key to achieving military operational objectives. These capabilities are further enhanced by operational flexibility and the ability to identify and take advantage of fleeting opportunities.

c. **Attack on Enemy COGs.** Attacks on enemy COGs typically continue during sustained operations. JFCs should time their actions to coincide with actions of other operations of the joint force and vice versa to achieve military strategic and operational-level objectives. As with all joint force operations, direct and indirect attacks of enemy COGs should be planned to achieve the required military strategic and operational-level objectives per the CONOPS, while limiting potential undesired effects on operations in follow-on phases.

d. **Synchronizing and/or Integrating Maneuver and Interdiction**

(1) Synchronizing and integrating air, land, maritime, and cyberspace interdiction and maneuver, enabled by JEMSO and space-based capabilities, provides one of the most dynamic concepts available to the joint force. Interdiction and maneuver usually are not considered separate operations against a common enemy, but rather are complementary operations planned to achieve the military strategic and operational-level objectives. Moreover, maneuver by air, land, or maritime forces can be conducted to interdict enemy military potential. Potential responses to integrated and synchronized maneuver and interdiction can create a dilemma for the enemy. If the enemy attempts to counter the maneuver, enemy forces may be exposed to unacceptable losses from interdiction. If the enemy employs measures to reduce such interdiction losses, enemy forces may not be able to counter the maneuver. The synergy achieved by integrating and synchronizing interdiction and maneuver throughout the OA assists commanders in optimizing leverage at the operational level.

(2) As a guiding principle, JFCs should exploit the flexibility inherent in joint force command relationships, joint targeting procedures, and other techniques to resolve the issues that can arise from the relationship between interdiction and maneuver. When interdiction and maneuver are employed, JFCs need to carefully balance the needs of surface maneuver forces, area-wide requirements for interdiction, and the undesirability of fragmenting joint force capabilities. The JFC's objectives, intent, and priorities, reflected in mission assignments and coordinating arrangements, enable subordinates to fully exploit the military potential of their forces while minimizing the friction generated by competing requirements. Effective targeting procedures in the joint force also alleviate such friction. As an example, interdiction requirements will often exceed interdiction means, requiring JFCs to prioritize requirements. Land and maritime force commanders responsible for integrating and synchronizing maneuver and interdiction within their AOs should be knowledgeable of JFC priorities and the responsibilities and authority assigned and delegated to commanders designated by the JFC to execute theater- and/or JOA-wide functions. JFCs alleviate this friction through the CONOPS and clear statements of intent for interdiction conducted relatively independent of surface maneuver operations. In doing this, JFCs rely on their vision as to how the major elements of the joint force contribute to achieving theater-strategic objectives. JFCs then employ a flexible range of techniques to

Large-Scale Combat Operations

assist in identifying requirements and applying capabilities to meet them. JFCs must define appropriate command relationships, establish effective joint targeting procedures, and make apportionment decisions.

(3) All commanders should consider how their operations can complement interdiction. These operations may include actions such as MILDEC, withdrawals, lateral repositioning, and flanking movements that are likely to cause the enemy to reposition surface forces, making them better targets for interdiction. Likewise, interdiction operations need to conform to and enhance the JFC's scheme of maneuver. This complementary use of maneuver and interdiction places the enemy in the operational dilemma of either defending from disadvantageous positions or exposing forces to interdiction strikes during attempted repositioning.

(4) Within the JOA, all joint force component operations must contribute to achievement of the JFC's objectives. To facilitate these operations, JFCs may establish AOs within their OA. **Synchronization and/or integration of maneuver and interdiction within land or maritime AOs is of particular importance,** particularly when JFCs task component commanders to execute theater- and/or JOA-wide functions.

(a) Air, land, and maritime commanders are directly concerned with those enemy forces and capabilities that can affect their current and future operations. Accordingly, that part of interdiction with a near-term effect on air, land, and maritime maneuver normally supports that maneuver. In fact, successful operations may depend on successful interdiction operations; for instance, to isolate the battle or weaken the enemy force before battle is fully joined.

(b) JFCs establish land and maritime AOs to decentralize execution of land and maritime component operations, allow rapid maneuver, and provide the ability to fight at extended ranges. The size, shape, and positioning of land or maritime AOs will be based on the JFC's CONOPS and the land or maritime commanders' requirements to accomplish their missions and protect their forces. **Within these AOs, land and maritime commanders are designated the supported commander for the integration and synchronization of maneuver, fires, and interdiction.** Accordingly, land and maritime commanders designate the target priority, effects, and timing of interdiction operations within their AOs. Further, in coordination with the land or maritime commander, a component commander designated as the supported commander for theater/JOA-wide interdiction has the latitude to plan and execute JFC prioritized missions within a land or maritime AO. If theater or JOA-wide interdiction operations would have adverse effects within a land or maritime AO, then the commander conducting those operations must either readjust the plan, resolve the issue with the appropriate component commander, or consult with the JFC for resolution.

(c) The land or maritime commander should clearly articulate the vision of maneuver operations to other commanders that may employ interdiction forces within the land or maritime AO. The land or maritime commander's intent and CONOPS should clearly state how interdiction will enable or enhance land or maritime force maneuver in the AO and what is to be accomplished with interdiction (as well as those actions to be

Chapter VIII

avoided, such as the destruction of key transportation nodes or the use of certain munitions in a specific area). Once this is understood, other interdiction-capable commanders can normally plan and execute operations with only that coordination required with the land or maritime commander. However, the land or maritime commander should provide other interdiction-capable commanders as much latitude as possible in the planning and execution of interdiction operations within the AO.

(d) Joint force operations in maritime or littoral OAs often require additional coordination among the maritime commander and other interdiction-capable commanders because of the highly specialized nature of some maritime operations, such as antisubmarine and mine warfare. This type of coordination requires that the interdiction-capable commanders maintain communication with the maritime commander. As in all operations, lack of close coordination among commanders in maritime OAs can result in friendly fire incidents and failed missions. The same principle applies concerning joint force air component mining operations in land or maritime OAs.

(5) JFCs need to pay particular attention and give priority to activities impinging on and supporting the maneuver and interdiction needs of all forces. In addition to normal target nomination procedures, JFCs establish procedures through which land or maritime force commanders can specifically identify those interdiction targets they are unable to engage with organic assets within their OAs that could affect planned or ongoing maneuver. These targets may be identified individually or by category, specified geographically, or tied to a desired effect or time period. Interdiction target priorities within the land or maritime OAs are considered along with theater and JOA-wide interdiction priorities by JFCs and reflected in the air apportionment decision. The JFACC uses these priorities to plan, coordinate, and execute the theater- and/or JOA-wide air interdiction effort. The purpose of these procedures is to afford added visibility to, and allow JFCs to give priority to, targets directly affecting planned maneuver by air, land, or maritime forces.

e. **Operations When WMD are Employed or Located**

(1) **Locating WMD and WMD Materials.** Since an enemy's use of WMD can quickly change the character of an operation or campaign, joint forces may be required to track, seize, and secure any WMD and materials used to develop WMD discovered or located in an OA. Once located, resources may be required to secure and inventory items for subsequent exploitation. If WMD sites are located, but joint forces are unable to seize and secure them, the JFC should plan to strike the sites if required to prevent WMD from being used or falling into enemy control. The desired effects of strikes are to minimize collateral effects and deny access to WMD. If sites are not under enemy control or in imminent jeopardy of falling to the enemy, monitor them persistently until sites can be seized and secured. During combat operations, exploitation, secure transport of WMD, and safe transport of technical personnel for disposition may depend upon a permissive OE.

(2) **Enemy Employment.** The use or the threatened use of WMD can cause large-scale shifts in strategic and operational-level objectives, phases, and COAs.

Multinational operations become more complicated with the threatened employment of these weapons. An enemy may use WMD against friendly force multinational partners, especially those with little or no defense against these weapons, to disintegrate the alliance or coalition. The enemy may also use chemical, biological or radiological weapons as part of an A2/AD plan.

(a) Intelligence and other joint staff members advise JFCs of an enemy's capability to employ WMD and under what conditions that enemy is most likely to do so. This advice includes an assessment of the enemy's willingness and intent to employ these weapons. It is important to ensure that high force or materiel concentrations do not provide lucrative targets for enemy WMD.

(b) Known threat of WMD use and associated preparedness against such use are imperative in this environment. The joint force can survive use of WMD by anticipating their employment and taking appropriate offensive and defensive measures. Commanders can protect their forces in a variety of ways, including training, MISO, OPSEC, dispersion of forces or materiel, use of IPE, and proper use of terrain for shielding against blast and radiation effects. Enhancement of CBRN defense capabilities may reduce incentives for a first strike by an enemy with WMD.

(c) The combination of active and passive defense can reduce the effectiveness or success of an enemy's use of WMD. The JFC may have to conduct offensive operations to control, defeat, disable, and/or dispose of enemy WMD capabilities before they can be brought to bear. Offensive measures include raids, strikes, and operations to locate and neutralize the threat of such weapons. When conducting offensive operations, the JFC must fully understand the collateral effects created by striking or neutralizing enemy WMD capabilities.

(d) JFCs should immediately inform HN authorities, the US ambassador or chief of mission, other USG departments and agencies, international organizations, and NGOs in the OA of enemy intentions to use WMD. These organizations do not have the same intelligence or decontamination capabilities as military units and need the maximum amount of time available to protect their personnel. In the event WMD are used against nonmilitary targets, JFCs must plan for and prepare to manage the consequences of CBRN incidents to mitigate the effects.

For additional guidance on defensive CBRN measures and countermeasures, refer to JP 3-11, Operations in Chemical, Biological, Radiological, and Nuclear Environments; *JP 3-40,* Countering Weapons of Mass Destruction; *JP 3-41,* Chemical, Biological, Radiological, and Nuclear Response; *and the* Department of Defense Strategy for Countering Weapons of Mass Destruction.

(3) **Friendly Employment.** When directed by the President and SecDef, CCDRs will plan for the employment of nuclear weapons by US forces in a manner consistent with national policy and strategic guidance. The employment of such weapons signifies an escalation of the war and is a presidential decision. USSTRATCOM's capabilities to lead in the collaborative planning of all nuclear missions are available to

support nuclear weapon employment. If directed to plan for the use of nuclear weapons, JFCs typically have two escalating objectives.

(a) The first objective is deterring or preventing the enemy from using WMD. Effective WMD deterrence rests on a credible deterrence policy that declares an adversary, expresses the will to pursue that adversary, and is backed by the capability to defend against the use and protect against the effects of WMD. A demonstrated collective military capability may contribute to the success of all three criteria for WMD deterrence. JFC deterrence efforts should involve security cooperation plans that emphasize the willingness of the US and its partners to employ forces for collective interests. Various bilateral and multilateral exercises and operations support deterrence by demonstrating collective willingness and capability to use force when necessary. Overall USG deterrence goals are supported by a credible capability to intercept WMD in transit; destroy critical nodes, links, and sources; defend against WMD attack; attribute WMD attacks; and dismantle WMD programs.

(b) If deterrence is not an effective option or fails, JFCs will respond appropriately, consistent with national policy and strategic guidance, to enemy aggression while seeking to control the intensity and scope of conflict and destruction. That response may include the employment of conventional, special operations, or nuclear forces.

(4) **CWMD.** JFC's should be prepared to conduct activities to curtail the conceptualization, development, possession, proliferation, use, and effects of WMD. When planning or executing operations and activities to counter WMD, JFCs coordinate and cooperate with not only other USG departments and agencies, but also local, tribal, and state organizations, in addition to multinational partners. With numerous stakeholders in the CWMD mission area, it is critical JFCs understand and consider the capabilities and responsibilities of various interorganizational partners when defining command relationships and coordinating interorganizational activities. Operations to counter WMD may require formation of a functional JTF for that purpose.

For further guidance on CWMD, refer to JP 3-40, Countering Weapons of Mass Destruction*.*

f. **Stability Activities.** Stability tasks and activities that began in previous phases may continue during this phase. These activities may focus on stability tasks that will help achieve strategic and operational-level objectives and create the conditions for the later stabilization and enable civil authority phases. Minimum essential stability activities should focus on protecting and facilitating the personal security and well-being of the civilian population. Stability activities provide minimum levels of security, food, water, shelter, and medical treatment. If no civilian or HN agency is present, capable, and willing, then JFCs and their staffs must resource these minimum essential stability tasks. When demand for resources exceeds the JFC's capability, higher level joint commanders should provide additional resources. These resources may be given to the requesting JFC or the mission may be given to follow-on forces to expeditiously conduct the tasks. JFCs at all levels assess resources available against the mission to determine how best to conduct these minimum essential stability tasks and what risk they can accept to accomplishment of combat tasks.

7. Considerations for Stabilization

a. **General.** Operations in a stabilize phase typically begin with significant military involvement, to include some combat and the potential for longer-term occupation. Operations then move increasingly toward transitioning to an interim civilian authority and enabling civil authority as the threat wanes and civil infrastructures are reestablished. The JFC's mission accomplishment requires fully integrating US military operations with the efforts of interorganizational participants in a comprehensive approach to accomplish assigned and implied tasks. As progress is made, military forces will increase their focus on supporting the efforts of HN authorities, other USG departments and agencies, international organizations, and/or NGOs. National Security Presidential Directive-44, *Management of Interagency Efforts Concerning Reconstruction and Stabilization,* assigns DOS the responsibility to plan and coordinate USG efforts in stabilization and reconstruction. The Secretary of State coordinates with SecDef to ensure harmonization with planned and ongoing operations. Military support to stabilization efforts within the JOA are the responsibility of the JFC.

For more information on stability activities, see JP 3-07, Stability.

b. **Several LOOs may be initiated immediately** (e.g., providing FHA, establishing security). In some cases, the scope of the problem set may dictate using other nonmilitary entities which are uniquely suited to address the problems. The goal of these military and civil efforts should be to eliminate root causes or deficiencies that create the problems (e.g., strengthen legitimate civil authority, rebuild government institutions, foster a sense of confidence and well-being, and support the conditions for economic reconstruction). With this in mind, the JFC may need to address how to coordinate CMO with the efforts of participating other government agencies, international organizations, NGOs, and HN assets.

For further guidance on CMO, refer to JP 3-57, Civil-Military Operations.

c. **Forces and Capabilities Mix.** The JFC may need to realign forces and capabilities or adjust force structure to begin stability activities in some portions of the OA even while sustained combat operations still are ongoing in other areas. For example, CA forces and HUMINT capabilities are critical to supporting stabilize phase operations and often involve a mix of forces and capabilities far different than those that supported the previous phases. Planning and continuous assessment will reveal the nature and scope of forces and capabilities required. These forces and capabilities may be available within the joint force or may be required from another theater or from the Reserve Component. The JFC should anticipate and request these forces and capabilities in a timely manner to facilitate their opportune employment.

d. **Stability Activities**

(1) As sustained combat operations conclude, military forces will shift their focus to stability activities as the military instrument's contribution to the more comprehensive **stabilization efforts** by all instruments of national power. Force protection will continue to be important, and combat operations might continue, although with less frequency and intensity than in the dominate phase. Of particular importance will be CMO, initially conducted to secure and safeguard the populace, reestablish civil law and order, protect or rebuild key infrastructure, and restore public services. US military forces should be prepared to lead the activities necessary to accomplish these tasks, especially if conducting a military occupation, and restore rule of law when indigenous civil, USG, multinational or international capacity does not exist or is incapable of assuming responsibility. Once legitimate civil authority is prepared to conduct such tasks, US military forces may support such activities as required/necessary. SFA plays an important part during stability activities by supporting and augmenting the development of the capacity and capability of FSFs and their supporting institutions. Likewise, the JFC's communication synchronization will play an important role in providing public information to foreign populations during this period.

For further guidance on SFA, refer to JP 3-22, Foreign Internal Defense.

(2) The military's predominant presence and its ability to command and control forces and logistics under extreme conditions may give it the de facto lead in stabilization efforts normally governed by other agencies that lack such capacities. However, most stability activities will likely be in support of, or transition to support of, US diplomatic, UN, or HN efforts. Integrated civilian and military efforts are key to success and military forces need to work competently in this environment while properly supporting the agency in charge. To be effective, planning and conducting stabilization efforts require a variety of perspectives and expertise and the cooperation and assistance of other USG departments and agencies, other Services, and alliance or multinational partners. Military forces should be prepared to work in integrated civilian-military teams that could include representatives from other US departments and agencies, foreign governments and security forces, international organizations, NGOs, and members of the private sector with relevant skills and expertise. Typical military support includes emergency infrastructure reconstruction, engineering, logistics, law enforcement, health services, and other activities to restore essential services.

For further guidance on stability activities and USG stabilization efforts, refer to JP 3-07, Stability, *and DODI 3000.05,* Stability Operations.

(a) CA forces are organized and trained to perform CA operations that support CMO conducted in conjunction with stability activities. MISO forces will develop, produce, and disseminate products to gain and reinforce popular support for the JFC's objectives. Complementing conventional forces' IW efforts, SOF will conduct FID to assess, train, advise, assist, and equip foreign military and paramilitary forces as they develop the capacity to secure their own lands and populations.

For further guidance on SOF, refer to JP 3-05, Special Operations.

(b) **CI activities** safeguard essential elements of friendly information. This is particularly pertinent in countering adversary HUMINT efforts. HN authorities, international organizations, and NGOs working closely with US forces may pass information (knowingly or unknowingly) to adversary elements that enables them to interfere with stability activities. Members of the local populace, who might actually be belligerents, often gain access to US military personnel and their bases by providing services such as laundry and cooking. They can then pass on information gleaned from that interaction to seek favor with a belligerent element or to avoid retaliation from belligerents. Identity activities, coupled with all-source intelligence analysis of the collected data, such as biometrics, forensics, signals intelligence, OSINT, and document and media exploitation, can support verification and deconfliction of HUMINT source identities and assist the JFC to take appropriate actions to counter potential compromise. CI personnel develop an estimate of the threat and recommend appropriate actions.

For more information on identity activities, see JDN 2-16, Identity Activities.

(c) **PA** activities support stability activities by providing public information about progress to internal and external audiences.

e. In the stabilize phase, commanders must consider **protection** from virtually any person, element, or group hostile to US interests. These could include activists, a group opposed to the operation, looters, and terrorists. Forces will have to be even more alert to force protection and security matters after a CBRN incident. JFCs also should be constantly ready to counter activity that could bring significant harm to units or jeopardize mission accomplishment. **Protection may involve the security of HN authorities, other USG department and agency personnel, and international organization and NGO members if authorized by higher authority.** For contractors, the GCC must evaluate the need for force protection support following the guidelines of DODI 3020.41, *Operational Contract Support (OCS)*.

f. Personnel should stay alert even in an operation with little or no perceived risk. **JFCs must take measures to prevent complacency and be ready to counter activity that could bring harm to units or jeopardize the operation.** However, security requirements should be balanced with the military operation's nature and objectives. During some stability activities, the use of certain security measures, such as carrying arms, wearing helmets and protective vests, or using secure communications may cause military forces to appear more threatening than intended, which may degrade the force's legitimacy and hurt relations with the local population.

g. **Restraint.** During the *stabilize* phase, military capability must be applied even more prudently since the support of the local population is essential for success. The actions of military personnel and units are framed by the disciplined application of force, including **specific ROE**. These ROE often will be more restrictive and detailed when compared to those for sustained combat operations due to national policy concerns. Moreover, these rules may change frequently during operations. Restraints on weaponry, tactics, and levels of violence characterize the environment. The use of excessive force could adversely affect efforts to gain or maintain legitimacy and impede the attainment of

Chapter VIII

both short- and long-term goals. The use of nonlethal capabilities should be considered to fill the gap between verbal warnings and deadly force when dealing with unarmed hostile elements and to avoid raising the level of conflict unnecessarily. The JFC must determine early in the planning stage what nonlethal technology is available, how well the force is trained to use it, and how the established ROE authorize its employment. The principle of restraint does not preclude the application of overwhelming force, when appropriate and authorized, to display US resolve and commitment. The reasons for the restraint often need to be understood by the individual Service member, because a single act could cause adverse diplomatic/political consequences.

h. **Perseverance.** Some operations may move quickly through the stabilize phase and transition smoothly to the enable civil authority phase. Other situations may require years of stabilization activities before this transition occurs. Therefore, the patient, resolute, and persistent pursuit for as long as necessary of the conditions desired to reach national objectives is often the requirement for success.

i. **Legitimacy.** Military activities must sustain the legitimacy of the operation and of the emerging or host government. During operations where a government does not exist, extreme caution should be used when dealing with individuals and organizations to avoid inadvertently legitimizing them. Implementation of strategic guidance through the CCS process can enhance perceptions of the legitimacy of stabilization efforts.

j. **OPSEC.** Although there may be no clearly defined threat, the essential elements of US military operations should be safeguarded. The uncertain nature of the situation, coupled with the potential for rapid change, requires that OPSEC be an integral part of all military operations. They can then pass on information gleaned from that interaction or provide other support to a belligerent element to seek favor or to avoid retaliation. The JFC must consider these and similar possibilities and take appropriate actions to counter potential compromise. OPSEC planners must consider the effect of media coverage and the possibility coverage may compromise essential security or disclose critical information.

8. Considerations for Enabling Civil Authority

a. **General.** In this phase, the joint operation is assessed and enabling objectives are established for transitioning from large-scale combat operations to FID and security cooperation. The catalyst for transition is that a legitimate civil authority has been established to manage the situation without further outside military intervention. The new government obtains legitimacy, and authority is transitioned from an interim civilian authority or transitional military authority to the new indigenous government. This situation may require a change in the joint operation as a result of an extension of the required stability activities in support of US diplomatic, HN, international organization, and/or NGO stabilization efforts.

b. **PB.** The transition from military operations to full civilian control may involve ongoing operations that have a significant combat component, including COIN operations, antiterrorism, and CT. Even while combat operations are ongoing, the operation will include a large stability component that is essentially a PB mission. PB, transitioning to a

DOS-led effort, provides the reconstruction and societal rehabilitation that offers hope to the HN populace. Stability measures establish the conditions that enable PB to succeed. PB promotes reconciliation, strengthens and rebuilds civil infrastructures and institutions, builds confidence, and supports economic reconstruction to prevent a return to conflict. The ultimate measure of success in PB is political, not military. Therefore, JFCs seek a clear understanding of the national/PN objectives and how military operations support that end state.

   c. **Transfer to Civil Authority.** In many cases, the US will transfer responsibility for the political and military affairs of the HN to another authority (e.g., UN observers, multinational peacekeeping force, or NATO) consistent with established termination criteria. This will probably occur after an extended period of conducting joint or multinational stability activities and PB missions as described above. Overall, transfer will likely occur in stages (e.g., HN sovereignty, PO under UN mandate, termination of all US military participation). Joint force support to this effort may include the following:

      (1) **Support to Truce Negotiations.** This support may include providing intelligence, security, transportation and other logistic support, and linguists for all participants.

      (2) **Transition to Civil Authority.** This transfer could be to a local or HN government or to an international authority facilitated by the UN. For example, an interim government (Northern Alliance) assumed governance during OEF and then transferred governance to a legitimate (newly elected) national government in Afghanistan. However, the coalition forces provided for the security of the nation, which was transferred later to the International Security Assistance Force as the authority for security that enabled the Afghan government. The transition for security is ongoing with the overall objective of transferring security concerns to the Afghan National Army, police, and other security forces.

   d. **Redeployment**

      (1) **Conduct.** Redeployment is the transfer of forces and materiel to support another JFC's operational requirements, or to return personnel, equipment, and materiel to home/demobilization stations for reintegration and out-processing. Redeployment is normally conducted in stages—the entire joint force will likely not redeploy in one relatively short period. It may include waste disposal, port operations, closing of contracts and other financial obligations, disposition of contracting records and files, clearing and marking of minefields and other explosive ordnance disposal activities, and ensuring appropriate units remain in place until their missions are complete. Redeployment must be planned and executed in a manner that facilitates the use of redeploying forces and supplies to meet new missions or crises.

         (a) Redeployment planning is the responsibility of the losing supported commander, when personnel, equipment, and materiel are redeployed to home or demobilization stations. The gaining supported commander is responsible for this planning when the redeployment is to a new OA.

Chapter VIII

(b) Upon redeployment, units or individuals may require refresher training prior to reassuming more traditional roles and missions. Service members and leaders may also require follow-on schooling to ensure normal career progression. Due to this, redeployment planning must be a collaborative and synchronized effort between supported and supporting commanders.

(2) **Redeployment to Other Contingencies.** Due to competing demands for limited forces, the joint force coordinator (for conventional forces) or joint force provider (for SOF and mobility forces) may source recommendations for allocating a force from one CCDR to another higher priority mission if the risks warrant. If SecDef approves the sourcing recommendation, the allocation will be ordered in a deployment order. Commanders and their staffs should consider how they would extricate forces and ensure they are prepared for the new contingency. This might include such things as a prioritized redeployment schedule, identification of aerial ports for linking intra- and intertheater airlift, the most recent intelligence assessments and supporting GEOINT products for the new contingency, and some consideration to achieving the national objectives of the original contingency through other means.

(3) **Redeployment in Support of Rotational Requirements.** Due to Service or other rotational requirements, forces may be relieved in place and redeployed to home station for reconstitution or regeneration. Commanders and their staffs must consider security and protective measures during the relief in place between incoming and outgoing forces.

For further information on redeployment, refer to JP 3-35, Deployment and Redeployment Operations, *and CJCSM 3130.06,* Global Force Management Allocation Policies and Procedures. *For further guidance on considerations for termination of operations, refer to JP 5-0,* Joint Planning, *and JP 3-33,* Joint Task Force Headquarters.

APPENDIX A
PRINCIPLES OF JOINT OPERATIONS

1. Introduction

The **principles of joint operations** are formed around the traditional **principles of war.** Three additional principles—restraint, perseverance, and legitimacy—are relevant to how the Armed Forces of the United States use combat power across the range of military operations. These three, added to the original nine, comprise 12 principles of joint operations. The principles do not apply equally in all joint operations. Most principles, if not all, are relevant in combat. Some principles, such as offensive, maneuver, and surprise, may not apply in some crisis response operations like FHA. However, principles such as unity of command, objective, and legitimacy are important in all operations.

2. Principles of Joint Operations

a. Objective

(1) The purpose of specifying the objective is to direct every military operation toward a clearly defined, decisive, and achievable goal.

(2) The purpose of military operations is to achieve specific objectives that support attainment of the overall strategic objectives identified to resolve the conflict. This frequently involves the destruction of the enemies' capabilities and their will to fight. The objective of joint operations not involving this destruction might be more difficult to define; nonetheless, it too must be clear from the beginning. Objectives must directly, quickly, and economically contribute to the purpose of the operation. Each operation must contribute to attaining strategic objectives. JFCs should avoid actions that do not contribute directly to achieving the objective(s).

(3) Additionally, changes to the military objectives may occur because national and military leaders gain a better understanding of the situation, or they may occur because the situation itself changes. **The JFC should anticipate these shifts in national objectives necessitating changes in the military objectives.** The changes may be very subtle, but if not made, achievement of the military objectives may no longer support the national objectives, legitimacy may be undermined, and force security may be compromised.

b. Offensive

(1) The purpose of an offensive action is to seize, retain, and exploit the initiative.

(2) Offensive action is the most effective and decisive way to achieve a clearly defined objective. Offensive operations are the means by which a military force seizes and holds the initiative while maintaining freedom of action and achieving decisive results. The importance of offensive action is fundamentally true across all levels of warfare.

Appendix A

(3) Commanders adopt the defensive only as a temporary expedient and must seek every opportunity to seize or regain the initiative. An offensive spirit must be inherent in the conduct of all defensive operations.

c. **Mass**

(1) The purpose of mass is to concentrate the effects of combat power at the most advantageous place and time to produce decisive results.

(2) In order to achieve mass, appropriate joint force capabilities are integrated and synchronized where they will have a decisive effect in a short period of time. Mass often must be sustained to have the desired effect. Massing effects of combat power, rather than concentrating forces, can enable even numerically inferior forces to produce decisive results and minimize human losses and waste of resources.

d. **Maneuver**

(1) The purpose of maneuver is to place the enemy in a position of disadvantage through the flexible application of combat power.

(2) Maneuver is the movement of forces in relation to the enemy to secure or retain positional advantage, usually in order to deliver—or threaten delivery of—the direct and indirect fires of the maneuvering force. Effective maneuver keeps the enemy off balance and thus also protects the friendly force. It contributes materially in exploiting successes, preserving freedom of action, and reducing vulnerability by continually posing new problems for the enemy.

e. **Economy of Force**

(1) The purpose of economy of force is to expend minimum essential combat power on secondary efforts in order to allocate the maximum possible combat power on primary efforts.

(2) Economy of force is the judicious employment and distribution of forces. It is the measured allocation of available combat power to such tasks as limited attacks, defense, delays, deception, or even retrograde operations to achieve mass elsewhere at the decisive point and time.

f. **Unity of Command**

(1) The purpose of unity of command is to ensure unity of effort under one responsible commander for every objective.

(2) Unity of command means that all forces operate under a single commander with the requisite authority to direct all forces employed in pursuit of a common purpose. Unity of command may not be possible during coordination and operations with multinational and interagency partners, but the requirement for unity of effort is paramount. Unity of effort—the coordination and cooperation toward common objectives, even if the

participants are not necessarily part of the same command or organization—is the product of successful unified action.

g. **Security**

(1) The purpose of security is to prevent the enemy from acquiring unexpected advantage.

(2) Security enhances freedom of action by reducing friendly vulnerability to hostile acts, influence, or surprise. Security results from the measures taken by commanders to protect their forces, the population, or other critical priorities. Staff planning and an understanding of enemy strategy, tactics, and doctrine enhance security. Risk is inherent in military operations. Application of this principle includes prudent risk management, not undue caution.

h. **Surprise**

(1) The purpose of surprise is to strike at a time or place or in a manner for which the enemy is unprepared.

(2) Surprise can help the commander shift the balance of combat power and thus achieve success well out of proportion to the effort expended. Factors contributing to surprise include speed in decision making, information sharing, and force movement; effective intelligence; deception; application of unexpected combat power; OPSEC; and variations in tactics and methods of operation.

i. **Simplicity**

(1) The purpose of simplicity is to increase the probability that plans and operations will be executed as intended by preparing clear, uncomplicated plans and concise orders.

(2) Simplicity contributes to successful operations. Simple plans and clear, concise orders minimize misunderstanding and confusion. When other factors are equal, the simplest plan is preferable. Simplicity in plans allows better understanding and execution planning at all echelons. Simplicity and clarity of expression greatly facilitate mission execution in the stress, fatigue, fog of war, and complexities of modern combat, and are especially critical to success in multinational operations.

j. **Restraint**

(1) The purpose of restraint is to prevent the unnecessary use of force.

(2) A single act could cause significant military and political consequences; therefore, judicious use of force is necessary. Restraint requires the careful and disciplined balancing of the need for security, the conduct of military operations, and national objectives. Excessive force antagonizes those parties involved, thereby damaging the legitimacy of the organization that uses it while potentially enhancing the legitimacy of the

Appendix A

opposing party. Sufficiently detailed ROE that the commander tailors to the specific circumstances of the operation can facilitate appropriate restraint.

k. Perseverance

(1) The purpose of perseverance is to ensure the commitment necessary to achieve national objectives.

(2) Perseverance involves preparation for measured, protracted military operations in pursuit of national objectives. Some joint operations may require years to reach the termination criteria. The underlying causes of the crisis may be elusive, making it difficult to achieve decisive resolution. The patient, resolute, and persistent pursuit of national goals and objectives often is essential to success. This will frequently involve diplomatic, economic, and informational measures to supplement military efforts.

l. Legitimacy

(1) The purpose of legitimacy is to maintain legal and moral authority in the conduct of operations.

(2) Legitimacy, which can be a decisive factor in operations, is based on the actual and perceived legality, morality, and rightness of the actions from the various perspectives of interested audiences. These audiences will include our national leadership and domestic population, governments, and civilian populations in the OA, and nations and organizations around the world.

(3) Committed forces must sustain the legitimacy of the operation and of the host government, where applicable. Security actions must be balanced with legitimacy concerns. All actions must be considered in the light of potentially competing strategic and tactical-level requirements, and must exhibit fairness in dealing with competing factions where appropriate. Legitimacy may depend on adherence to objectives agreed to by the international community, ensuring the action is appropriate to the situation and to perceptions of fairness in dealing with various factions. Restricting the use of force, restructuring the type of forces employed, protecting civilians, and ensuring the disciplined conduct of the forces involved may reinforce legitimacy.

(4) Another aspect of this principle is the legitimacy bestowed upon a local government through the perception of the populace that it governs. Humanitarian and civil military operations help develop a sense of legitimacy for the supported government. Because the populace perceives that the government has genuine authority to govern and uses proper agencies for valid purposes, they consider that government as legitimate, especially when coupled with successful efforts to build the capability and capacity of the supported government to complete such operations on its own. During operations in an area where a legitimate government does not exist, extreme caution should be used when dealing with individuals and organizations to avoid inadvertently legitimizing them.

APPENDIX B
REFERENCES

The development of JP 3-0 is based upon the following primary references:

1. General

   a. The Goldwater-Nichols Department of Defense Reorganization Act of 1986 (Title 10, USC, Section 161).

   b. *The National Security Strategy of the United States of America.*

   c. *National Defense Strategy of the United States of America.*

   d. *National Defense Authorization Act of 1991.*

   e. *National Military Strategy.*

   f. *Department of Defense Strategy for Countering Weapons of Mass Destruction.*

   g. *The Department of Defense Cyber Strategy.*

   h. *National Strategy for Homeland Security.*

   i. *National Response Framework.*

   j. *The Quadrennial Defense Review 2014.*

   k. *Unified Command Plan.*

   l. *Guidance for Employment of the Force.*

   m. *Defense Planning Guidance.*

   n. *Joint Strategic Capabilities Plan.*

   o. Executive Order 12656, *Assignment of Emergency Preparedness Responsibilities.*

   p. National Security Presidential Directive-44, *Management of Interagency Efforts Concerning Reconstruction and Stabilization.*

   q. *UN Charter.*

   r. Executive Order 12333, *United States Intelligence Activities.*

2. Department of Defense Publications

   a. DOD 5240.1-R, *Procedures Governing the Activities of DOD Intelligence Components that Affect United States Persons.*

Appendix B

b. DODD 3000.03E, *DOD Executive Agent for Non-Lethal Weapons (NLW), and NLW Policy.*

c. DODD 3025.18, *Defense Support of Civil Authorities (DSCA).*

d. DODD 3160.01, *Homeland Defense Activities Conducted by the National Guard.*

e. DODD 5240.01, *DOD Intelligence Activities.*

f. DODI 3020.41, *Operational Contract Support (OCS).*

g. DODI 3025.21, *Defense Support of Civilian Law Enforcement Agencies.*

h. DODI, 3025.22, *The Use of the National Guard for Defense Support of Civil Authorities.*

i. DODI 4715.19, *Use of Open-Air Burn Pits in Contingency Operations.*

j. DODI 4715.22, *Environmental Management Policy for Contingency Locations.*

k. DODI 6490.03, *Deployment Health.*

l. DODI 8500.01, *Cybersecurity.*

3. Chairman of the Joint Chiefs of Staff Publications

a. CJCSI 3121.01B, *Standing Rules of Engagement/Standing Rules for the Use of Force for US Forces.*

b. CJCSI 3126.01A, *Language, Regional Expertise, and Culture (LREC) Capability Identification, Planning, and Sourcing.*

c. CJCSI 3141.01E, *Management and Review of Joint Strategic Capabilities Plan (JSCP)-Tasked Plans.*

d. CJCSI 3150.25F, *Joint Lessons Learned Program.*

e. CJCSI 3320.01D, *Joint Electromagnetic Spectrum Operations (JEMSO).*

f. CJCSI 3500.01H, *Joint Training Policy for the Armed Forces of the United States.*

g. CJCSI 5715.01C, *Joint Staff Participation in Interagency Affairs.*

h. CJCSI 5810.01D, *Implementation of the DOD Law of War Program.*

i. CJCSM 3122.01A, *Joint Operation Planning and Execution System (JOPES) Volume I (Planning Policies and Procedures).*

j. CJCSM 3122.05, *Operating Procedures for Joint Operation Planning and Execution System (JOPES)-Information Systems (IS) Governance.*

k. CJCSM 3130.01A, *Campaign Planning Procedures and Responsibilities.*

l. CJCSM 3130.03, *Adaptive Planning and Execution (APEX) Planning Formats and Guidance.*

m. CJCSM 3150.03D, *Joint Reporting Structure Event and Incident Reports.*

n. JP 1, *Doctrine for the Armed Forces of the United States.*

o. JP 1-0, *Joint Personnel Support.*

p. JP 2-0, *Joint Intelligence.*

q. JP 2-01, *Joint and National Intelligence Support to Military Operations.*

r. JP 2-01.2, *Counterintelligence and Human Intelligence in Joint Operations.*

s. JP 2-01.3, *Joint Intelligence Preparation of the Operational Environment.*

t. JP 3-01, *Countering Air and Missile Threats.*

u. JP 3-02, *Amphibious Operations.*

v. JP 3-03, *Joint Interdiction.*

w. JP 3-05, *Special Operations.*

x. JP 3-05.1, *Unconventional Warfare.*

y JP 3-07, *Stability.*

z. JP 3-07.3, *Peace Operations.*

aa. JP 3-07.4, *Counterdrug Operations.*

bb. JP 3-08, *Interorganizational Cooperation.*

cc. JP 3-09, *Joint Fire Support.*

dd. JP 3-09.3, *Close Air Support.*

ee. JP 3-10, *Joint Security Operations in Theater.*

ff. JP 3-11, *Operations in Chemical, Biological, Radiological, and Nuclear Environments.*

Appendix B

gg. JP 3-12, *Cyberspace Operations.*

hh. JP 3-13, *Information Operations.*

ii. JP 3-13.1, *Electronic Warfare.*

jj. JP 3-13.2, *Military Information Support Operations.*

kk. JP 3-13.4, *Military Deception.*

ll. JP 3-14, *Space Operations.*

mm. JP 3-15.1, *Counter-Improvised Explosive Device Operations.*

nn. JP 3-16, *Multinational Operations.*

oo. JP 3-17, *Air Mobility Operations.*

pp. JP 3-18, *Joint Forcible Entry Operations.*

qq. JP 3-20, *Security Cooperation.*

rr. JP 3-22, *Foreign Internal Defense.*

ss. JP 3-24, *Counterinsurgency.*

tt. JP 3-26, *Counterterrorism.*

uu. JP 3-27, *Homeland Defense.*

vv. JP 3-28, *Defense Support of Civil Authorities.*

ww. JP 3-29, *Foreign Humanitarian Assistance.*

xx. JP 3-30, *Command and Control of Joint Air Operations.*

yy. JP 3-31, *Command and Control for Joint Land Operations.*

zz. JP 3-32, *Command and Control for Joint Maritime Operations.*

aaa. JP 3-33, *Joint Task Force Headquarters.*

bbb. JP 3-34, *Joint Engineer Operations.*

ccc. JP 3-40, *Countering Weapons of Mass Destruction.*

ddd. JP 3-41, *Chemical, Biological, Radiological, and Nuclear Response.*

eee. JP 3-57, *Civil-Military Operations.*

fff. JP 3-59, *Meteorological and Oceanographic Operations.*

ggg. JP 3-60, *Joint Targeting.*

hhh. JP 3-61, *Public Affairs.*

iii. JP 3-68, *Noncombatant Evacuation Operations.*

jjj. JP 4-0, *Joint Logistics.*

kkk. JP 4-09, *Distribution Operations.*

lll. JP 4-10, *Operational Contract Support.*

mmm. JP 5-0, *Joint Planning.*

nnn. JP 6-0, *Joint Communications System.*

ooo. JP 6-01, *Joint Electromagnetic Spectrum Management Operations.*

4. Allied Joint Publications

a. AJP-01(D), *Allied Joint Doctrine.*

b. AJP-3(B), *Allied Joint Doctrine for the Conduct of Operations.*

Intentionally Blank

APPENDIX C
ADMINISTRATIVE INSTRUCTIONS

1. User Comments

Users in the field are highly encouraged to submit comments on this publication to: Joint Staff J-7, Deputy Director, Joint Education and Doctrine, ATTN: Joint Doctrine Analysis Division, 116 Lake View Parkway, Suffolk, VA 23435-2697. These comments should address content (accuracy, usefulness, consistency, and organization), writing, and appearance.

2. Authorship

The lead agent and Joint Staff doctrine sponsor for this publication is the Joint Staff Director for Joint Force Development (J-7).

3. Supersession

This publication supersedes JP 3-0, *Joint Operations*, 11 August 2011.

4. Change Recommendations

a. Recommendations for urgent changes to this publication should be submitted:

TO: Deputy Director, Joint Education and Doctrine (DD JED), Attn: Joint Doctrine Division, 7000 Joint Staff (J-7), Washington, DC, 20318-7000 or email:js.pentagon.j7.list.dd-je-d-jdd-all@mail.mil.

b. Routine changes should be submitted electronically to the Deputy Director, Joint Education and Doctrine, ATTN: Joint Doctrine Analysis Division, 116 Lake View Parkway, Suffolk, VA 23435-2697, and info the lead agent and the Director for Joint Force Development, J-7/JED.

c. When a Joint Staff directorate submits a proposal to the CJCS that would change source document information reflected in this publication, that directorate will include a proposed change to this publication as an enclosure to its proposal. The Services and other organizations are requested to notify the Joint Staff J-7 when changes to source documents reflected in this publication are initiated.

5. Lessons Learned

The Joint Lessons Learned Program (JLLP) primary objective is to enhance joint force readiness and effectiveness by contributing to improvements in doctrine, organization, training, materiel, leadership and education, personnel, facilities, and policy. The Joint Lessons Learned Information System (JLLIS) is the DOD system of record for lessons learned and facilitates the collection, tracking, management, sharing, collaborative resolution, and dissemination of lessons learned to improve the development and readiness of the joint force. The JLLP integrates with joint doctrine through the joint doctrine

Appendix C

development process by providing lessons and lessons learned derived from operations, events, and exercises. As these inputs are incorporated into joint doctrine, they become institutionalized for future use, a major goal of the JLLP. Lessons and lessons learned are routinely sought and incorporated into draft JPs throughout formal staffing of the development process. The JLLIS Website can be found at https://www.jllis.mil or http://www.jllis.smil.mil.

6. Distribution of Publications

Local reproduction is authorized, and access to unclassified publications is unrestricted. However, access to and reproduction authorization for classified JPs must be IAW DOD Manual 5200.01, Volume 1, *DOD Information Security Program: Overview, Classification, and Declassification,* and DOD Manual 5200.01, Volume 3, *DOD Information Security Program: Protection of Classified Information.*

7. Distribution of Electronic Publications

a. Joint Staff J-7 will not print copies of JPs for distribution. Electronic versions are available on JDEIS Joint Electronic Library Plus (JEL+) at https://jdeis.js.mil/jdeis/index.jsp (NIPRNET) and http://jdeis.js.smil.mil/jdeis/index.jsp (SIPRNET), and on the JEL at http://www.dtic.mil/doctrine (NIPRNET).

b. Only approved JPs are releasable outside the combatant commands, Services, and Joint Staff. Defense attachés may request classified JPs by sending written requests to Defense Intelligence Agency (DIA)/IE-3, 200 MacDill Blvd., Joint Base Anacostia-Bolling, Washington, DC 20340-5100.

c. JEL CD-ROM. Upon request of a joint doctrine development community member, the Joint Staff J-7 will produce and deliver one CD-ROM with current JPs. This JEL CD-ROM will be updated not less than semi-annually and when received can be locally reproduced for use within the combatant commands, Services, and combat support agencies.

GLOSSARY
PART I—ABBREVIATIONS, ACRONYMS, AND INITIALISMS

A2	antiaccess
AADC	area air defense commander
ACA	airspace control authority
ACO	airspace control order
ACP	airspace control plan
ACS	airspace control system
AD	area denial
AJP	Allied joint publication
AO	area of operations
AOA	amphibious objective area
AOR	area of responsibility
APEX	Adaptive Planning and Execution
C2	command and control
CA	civil affairs
CBRN	chemical, biological, radiological, and nuclear
CCDR	combatant commander
CCIR	commander's critical information requirement
CCMD	combatant command
CCS	commander's communication synchronization
CD	counterdrug
CDRUSSOCOM	Commander, United States Special Operations Command
CDRUSSTRATCOM	Commander, United States Strategic Command
CI	counterintelligence
CID	combat identification
C-IED	counter-improvised explosive device
CJCS	Chairman of the Joint Chiefs of Staff
CJCSI	Chairman of the Joint Chiefs of Staff instruction
CJCSM	Chairman of the Joint Chiefs of Staff manual
CJTF	commander, joint task force
CMO	civil-military operations
CMOC	civil-military operations center
CO	cyberspace operations
COA	course of action
COCOM	combatant command (command authority)
COG	center of gravity
COIN	counterinsurgency
CONOPS	concept of operations
CONUS	continental United States
COP	common operational picture
CSE	cyberspace support element
CT	counterterrorism
CTF	counter threat finance

Glossary

CWMD	countering weapons of mass destruction
DAFL	directive authority for logistics
DCA	defensive counterair
DCO	defensive cyberspace operations
DHS	Department of Homeland Security
DLA	Defense Logistics Agency
DOD	Department of Defense
DODD	Department of Defense directive
DODI	Department of Defense instruction
DODIN	Department of Defense information network
DOS	Department of State
DSCA	defense support of civil authorities
DSPD	defense support to public diplomacy
DSR	Defense Strategy Review
EA	electronic attack
EMBM	electromagnetic battle management
EME	electromagnetic environment
EMS	electromagnetic spectrum
EW	electronic warfare
EWCA	electronic warfare control authority
FCC	functional combatant commander
FCP	functional campaign plan
FDO	flexible deterrent option
FFIR	friendly force information requirement
FHA	foreign humanitarian assistance
FHP	force health protection
FID	foreign internal defense
FLOT	forward line of own troops
FRO	flexible response option
FSF	foreign security forces
GCC	geographic combatant commander
GEF	Guidance for Employment of the Force
GEOINT	geospatial intelligence
HD	homeland defense
HN	host nation
HNS	host-nation support
HQ	headquarters
HUMINT	human intelligence
I2	identity intelligence
IAW	in accordance with

Glossary

IED	improvised explosive device
IM	information management
IO	information operations
IPE	individual protective equipment
IRC	information-related capability
ISR	intelligence, surveillance, and reconnaissance
IW	irregular warfare
J-2	intelligence directorate of a joint staff
J-3	operations directorate of a joint staff
J-5	plans directorate of a joint staff
J-7	training directorate of a joint staff
JDN	joint doctrine note
JEMSO	joint electromagnetic spectrum operations
JFACC	joint force air component commander
JFC	joint force commander
JFLCC	joint force land component commander
JFMCC	joint force maritime component commander
JFSOCC	joint force special operations component commander
JIACG	joint interagency coordination group
JIPOE	joint intelligence preparation of the operational environment
JOA	joint operations area
JP	joint publication
JPP	joint planning process
JRSOI	joint reception, staging, onward movement, and integration
JSA	joint security area
JSCP	Joint Strategic Capabilities Plan
JSOA	joint special operations area
JTCB	joint targeting coordination board
JTF	joint task force
JUO	joint urban operation
KLE	key leader engagement
LEA	law enforcement agency
LNO	liaison officer
LOC	line of communications
LOE	line of effort
LOO	line of operation
MILDEC	military deception
MIPOE	medical intelligence preparation of the operational environment
MISO	military information support operations

MOE	measure of effectiveness
MOP	measure of performance
NATO	North Atlantic Treaty Organization
NEO	noncombatant evacuation operation
NGA	National Geospatial-Intelligence Agency
NGO	nongovernmental organization
NMS	national military strategy
NRF	National Response Framework
NSC	National Security Council
NSS	national security strategy
OA	operational area
OCA	offensive counterair
OCS	operational contract support
OE	operational environment
OEF	Operation ENDURING FREEDOM
OIF	Operation IRAQI FREEDOM
OPCON	operational control
OPLAN	operation plan
OPORD	operation order
OPSEC	operations security
OSINT	open-source intelligence
PA	public affairs
PB	peace building
PEO	peace enforcement operations
PIR	priority intelligence requirement
PKO	peacekeeping operations
PM	peacemaking
PMESII	political, military, economic, social, information, and infrastructure
PN	partner nation
PNT	positioning, navigation, and timing
PO	peace operations
PR	personnel recovery
ROE	rules of engagement
SCA	space coordinating authority
SecDef	Secretary of Defense
SFA	security force assistance
SJA	staff judge advocate
SOF	special operations forces
SOP	standard operating procedure

Glossary

TACON	tactical control
TCP	theater campaign plan
TDP	theater distribution plan
TF	task force
TLO	theater logistics overview
TMM	transregional, multi-domain, and multi-functional
TSOC	theater special operations command
UCP	Unified Command Plan
UN	United Nations
USAID	United States Agency for International Development
USC	United States Code
USCYBERCOM	United States Cyber Command
USG	United States Government
USSTRATCOM	United States Strategic Command
USTRANSCOM	United States Transportation Command
UW	unconventional warfare
WMD	weapons of mass destruction

PART II—TERMS AND DEFINITIONS

activity. 1. A unit, organization, or installation performing a function or mission. 2. A function, mission, action, or collection of actions. (DOD Dictionary. SOURCE: JP 3-0)

adversary. A party acknowledged as potentially hostile to a friendly party and against which the use of force may be envisaged. (DOD Dictionary. SOURCE: JP 3-0)

air apportionment. The determination and assignment of the total expected effort by percentage and/or by priority that should be devoted to the various air operations for a given period of time. (DOD Dictionary. SOURCE: JP 3-0)

alliance. The relationship that results from a formal agreement between two or more nations for broad, long-term objectives that further the common interests of the members. (DOD Dictionary. SOURCE: JP 3-0)

antiaccess. Action, activity, or capability, usually long-range, designed to prevent an advancing enemy force from entering an operational area. Also called **A2**. (Approved for inclusion in the DOD Dictionary.)

area denial. Action, activity, or capability, usually short-range, designed to limit an enemy force's freedom of action within an operational area. Also called **AD**. (Approved for inclusion in the DOD Dictionary.)

area of influence. A geographical area wherein a commander is directly capable of influencing operations by maneuver or fire support systems normally under the commander's command or control. (DOD Dictionary. SOURCE: JP 3-0)

area of interest. That area of concern to the commander, including the area of influence, areas adjacent thereto, and extending into enemy territory. Also called **AOI**. (Approved for incorporation into the DOD Dictionary.)

area of operations. An operational area defined by a commander for land and maritime forces that should be large enough to accomplish their missions and protect their forces. Also called **AO**. (Approved for incorporation into the DOD Dictionary.)

assessment. 1. A continuous process that measures the overall effectiveness of employing capabilities during military operations. 2. Determination of the progress toward accomplishing a task, creating a condition, or achieving an objective. 3. Analysis of the security, effectiveness, and potential of an existing or planned intelligence activity. 4. Judgment of the motives, qualifications, and characteristics of present or prospective employees or "agents." (Approved for incorporation into the DOD Dictionary.)

assign. 1. To place units or personnel in an organization where such placement is relatively permanent, and/or where such organization controls and administers the units or personnel for the primary function, or greater portion of the functions, of the unit or personnel. 2. To detail individuals to specific duties or functions where such duties or

Glossary

functions are primary and/or relatively permanent. (DOD Dictionary. SOURCE: JP 3-0)

attach. 1. The placement of units or personnel in an organization where such placement is relatively temporary. 2. The detailing of individuals to specific functions where such functions are secondary or relatively temporary. (DOD Dictionary. SOURCE: JP 3-0)

battle damage assessment. The estimate of damage composed of physical and functional damage assessment, as well as target system assessment, resulting from the application of lethal or nonlethal military force. Also called **BDA.** (DOD Dictionary. SOURCE: JP 3-0)

boundary. A line that delineates surface areas for the purpose of facilitating coordination and deconfliction of operations between adjacent units, formations, or areas. (DOD Dictionary. SOURCE: JP 3-0)

close air support. Air action by manned or unmanned fixed-wing and rotary-wing aircraft against hostile targets that are in close proximity to friendly forces and that require detailed integration of each air mission with the fire and movement of those forces. Also called **CAS.** (Approved for incorporation into the DOD Dictionary.)

combatant commander. A commander of one of the unified or specified combatant commands established by the President. Also called **CCDR.** (DOD Dictionary. SOURCE: JP 3-0)

combat power. The total means of destructive and/or disruptive force that a military unit/formation can apply against the opponent at a given time. (Approved for incorporation into the DOD Dictionary.)

commander's critical information requirement. An information requirement identified by the commander as being critical to facilitating timely decision making. Also called **CCIR.** (DOD Dictionary. SOURCE: JP 3-0)

commander's intent. A clear and concise expression of the purpose of the operation and the desired military end state that supports mission command, provides focus to the staff, and helps subordinate and supporting commanders act to achieve the commander's desired results without further orders, even when the operation does not unfold as planned. (DOD Dictionary. SOURCE: JP 3-0)

command post exercise. None. (Approved for removal from the DOD Dictionary.)

common operational picture. A single identical display of relevant information shared by more than one command that facilitates collaborative planning and assists all echelons to achieve situational awareness. Also called **COP.** (DOD Dictionary. SOURCE: JP 3-0)

Glossary

condition. 1. Those variables of an operational environment or situation in which a unit, system, or individual is expected to operate and may affect performance. 2. A physical or behavioral state of a system that is required for the achievement of an objective. (DOD Dictionary. SOURCE: JP 3-0)

continuity of operations. The degree or state of being continuous in the conduct of functions, tasks, or duties necessary to accomplish a military action or mission in carrying out the national military strategy. Also called **COOP**. (DOD Dictionary. SOURCE: JP 3-0)

control. 1. Authority that may be less than full command exercised by a commander over part of the activities of subordinate or other organizations. (JP 1) 2. In mapping, charting, and photogrammetry, a collective term for a system of marks or objects on the Earth or on a map or a photograph, whose positions or elevations (or both) have been or will be determined. (JP 2-03) 3. Physical or psychological pressures exerted with the intent to assure that an agent or group will respond as directed. (JP 3-0) 4. In intelligence usage, an indicator governing the distribution and use of documents, information, or material. (Approved for incorporation into the DOD Dictionary.)

coup de main. None. (Approved for removal from the DOD Dictionary.)

crisis. An incident or situation involving a threat to the United States, its citizens, military forces, or vital interests that develops rapidly and creates a condition of such diplomatic, economic, or military importance that commitment of military forces and resources is contemplated to achieve national objectives. (DOD Dictionary. SOURCE: JP 3-0)

cyberspace operations. The employment of cyberspace capabilities where the primary purpose is to achieve objectives in or through cyberspace. Also called **CO**. (Approved for incorporation into the DOD Dictionary.)

deterrence. The prevention of action by the existence of a credible threat of unacceptable counteraction and/or belief that the cost of action outweighs the perceived benefits. (DOD Dictionary. SOURCE: JP 3-0)

disarmament. None. (Approved for removal from the DOD Dictionary.)

economy of force. The judicious employment and distribution of forces so as to expend the minimum essential combat power on secondary efforts in order to allocate the maximum possible combat power on primary efforts. (DOD Dictionary. SOURCE: JP 3-0)

effect. 1. The physical or behavioral state of a system that results from an action, a set of actions, or another effect. 2. The result, outcome, or consequence of an action. 3. A change to a condition, behavior, or degree of freedom. (DOD Dictionary. SOURCE: JP 3-0)

end state. The set of required conditions that defines achievement of the commander's objectives. (DOD Dictionary. SOURCE: JP 3-0)

engagement. 1. In air defense, an attack with guns or air-to-air missiles by an interceptor aircraft, or the launch of an air defense missile by air defense artillery and the missile's subsequent travel to intercept. (JP 3-01) 2. A tactical conflict, usually between opposing lower echelons maneuver forces. (DOD Dictionary. SOURCE: JP 3-0)

essential element of friendly information. Key question likely to be asked by adversary officials and intelligence systems about specific friendly intentions, capabilities, and activities, so they can obtain answers critical to their operational effectiveness. Also called **EEFI**. (Approved for replacement of "essential elements of friendly information" and its definition in the DOD Dictionary.)

exclusion zone. A zone established by a sanctioning body to prohibit specific activities in a specific geographic area in order to persuade nations or groups to modify their behavior to meet the desires of the sanctioning body or face continued imposition of sanctions, or the use or threat of force. (DOD Dictionary. SOURCE: JP 3-0)

exercise. A military maneuver or simulated wartime operation involving planning, preparation, and execution that is carried out for the purpose of training and evaluation. (Approved for incorporation into the DOD Dictionary with JP 3-0 as the source JP.)

expeditionary force. An armed force organized to achieve a specific objective in a foreign country. (Approved for incorporation into the DOD Dictionary.)

fire support coordination measure. A measure employed by commanders to facilitate the rapid engagement of targets and simultaneously provide safeguards for friendly forces. Also called **FSCM**. (DOD Dictionary. SOURCE: JP 3-0)

force projection. The ability to project the military instrument of national power from the United States or another theater, in response to requirements for military operations. (DOD Dictionary. SOURCE: JP 3-0)

force protection. Preventive measures taken to mitigate hostile actions against Department of Defense personnel (to include family members), resources, facilities, and critical information. Also called **FP**. (DOD Dictionary. SOURCE: JP 3-0)

foreign assistance. Assistance to foreign nations ranging from the sale of military equipment and support for foreign internal defense to donations of food and medical supplies to aid survivors of natural and man-made disasters that may be provided through development assistance, humanitarian assistance, and security assistance. (Approved for incorporation into the DOD Dictionary.)

freedom of navigation operations. Operations conducted to protect United States navigation, overflight, and related interests on, under, and over the seas. (Approved for incorporation into the DOD Dictionary.)

Glossary

friendly force information requirement. Information the commander and staff need to understand the status of friendly force and supporting capabilities. Also called **FFIR**. (DOD Dictionary. SOURCE: JP 3-0)

full-spectrum superiority. The cumulative effect of dominance in the air, land, maritime, and space domains, electromagnetic spectrum, and information environment (which includes cyberspace) that permits the conduct of joint operations without effective opposition or prohibitive interference. (Approved for incorporation into the DOD Dictionary.)

hostile environment. Operational environment in which host government forces, whether opposed to or receptive to operations that a unit intends to conduct, do not have control of the territory and population in the intended operational area. (Approved for inclusion in the DOD Dictionary.)

information management. The function of managing an organization's information resources for the handling of data and information acquired by one or many different systems, individuals, and organizations in a way that optimizes access by all who have a share in that data or a right to that information. Also called **IM**. (DOD Dictionary. SOURCE: JP 3-0)

interagency coordination. Within the context of Department of Defense involvement, the coordination that occurs between elements of Department of Defense, and participating United States Government departments and agencies for the purpose of achieving an objective. (Approved for incorporation into the DOD Dictionary.)

interoperability. 1. The ability to act together coherently, effectively, and efficiently to achieve tactical, operational, and strategic objectives. (JP 3-0) 2. The condition achieved among communications-electronics systems or items of communications-electronics equipment when information or services can be exchanged directly and satisfactorily between them and/or their users. (JP 6-0) (Approved for incorporation into the DOD Dictionary.)

joint fires. Fires delivered during the employment of forces from two or more components in coordinated action to produce desired effects in support of a common objective. (DOD Dictionary. SOURCE: JP 3-0)

joint fire support. Joint fires that assist air, land, maritime, and special operations forces to move, maneuver, and control territory, populations, airspace, and key waters. (DOD Dictionary. SOURCE: JP 3-0)

joint force. A force composed of elements, assigned or attached, of two or more Military Departments operating under a single joint force commander. (Approved for incorporation into the DOD Dictionary.)

joint force air component commander. The commander within a unified command, subordinate unified command, or joint task force responsible to the establishing commander for recommending the proper employment of assigned, attached, and/or made available for tasking air forces; planning and coordinating air operations; or accomplishing such operational missions as may be assigned. Also called **JFACC.** (DOD Dictionary. SOURCE: JP 3-0)

joint force land component commander. The commander within a unified command, subordinate unified command, or joint task force responsible to the establishing commander for recommending the proper employment of assigned, attached, and/or made available for tasking land forces; planning and coordinating land operations; or accomplishing such operational missions as may be assigned. Also called **JFLCC.** (DOD Dictionary. SOURCE: JP 3-0)

joint force maritime component commander. The commander within a unified command, subordinate unified command, or joint task force responsible to the establishing commander for recommending the proper employment of assigned, attached, and/or made available for tasking maritime forces and assets; planning and coordinating maritime operations; or accomplishing such operational missions as may be assigned. Also called **JFMCC.** (DOD Dictionary. SOURCE: JP 3-0)

joint force special operations component commander. The commander within a unified command, subordinate unified command, or joint task force responsible to the establishing commander for recommending the proper employment of assigned, attached, and/or made available for tasking special operations forces and assets; planning and coordinating special operations; or accomplishing such operational missions as may be assigned. Also called **JFSOCC.** (DOD Dictionary. SOURCE: JP 3-0)

joint functions. Related capabilities and activities placed into six basic groups of command and control, intelligence, fires, movement and maneuver, protection, and sustainment to help joint force commanders synchronize, integrate, and direct joint operations. (DOD Dictionary. SOURCE: JP 3-0)

joint operations. Military actions conducted by joint forces and those Service forces employed in specified command relationships with each other, which of themselves, do not establish joint forces. (Approved for incorporation into the DOD Dictionary.)

joint operations area. An area of land, sea, and airspace, defined by a geographic combatant commander or subordinate unified commander, in which a joint force commander (normally a joint task force commander) conducts military operations to accomplish a specific mission. Also called **JOA.** (DOD Dictionary. SOURCE: JP 3-0)

joint special operations area. An area of land, sea, and airspace assigned by a joint force commander to the commander of a joint special operations force to conduct special operations activities. Also called **JSOA.** (DOD Dictionary. SOURCE: JP 3-0)

link. 1. A behavioral, physical, or functional relationship between nodes. 2. In communications, a general term used to indicate the existence of communications facilities between two points. 3. A maritime route, other than a coastal or transit route that connects any two or more routes together. (Approved for incorporation into the DOD Dictionary.)

major operation. 1. A series of tactical actions (battles, engagements, strikes) conducted by combat forces, coordinated in time and place, to achieve strategic or operational objectives in an operational area. 2. For noncombat operations, a reference to the relative size and scope of a military operation. (Approved for incorporation into the DOD Dictionary.)

maneuver. 1. A movement to place ships, aircraft, or land forces in a position of advantage over the enemy. 2. A tactical exercise carried out at sea, in the air, on the ground, or on a map in imitation of war. 3. The operation of a ship, aircraft, or vehicle, to cause it to perform desired movements. 4. Employment of forces in the operational area through movement in combination with fires to achieve a position of advantage in respect to the enemy. (DOD Dictionary. SOURCE: JP 3-0)

military engagement. Routine contact and interaction between individuals or elements of the Armed Forces of the United States and those of another nation's armed forces, or foreign and domestic civilian authorities or agencies to build trust and confidence, share information, coordinate mutual activities, and maintain influence. (DOD Dictionary. SOURCE: JP 3-0)

military intervention. The deliberate act of a nation or a group of nations to introduce its military forces into the course of an existing controversy. (DOD Dictionary. SOURCE: JP 3-0)

military occupation. A condition in which territory is under the effective control of a foreign armed force. (DOD Dictionary. SOURCE: JP 3-0)

mission. 1. The task, together with the purpose, that clearly indicates the action to be taken and the reason therefore. (JP 3-0) 2. In common usage, especially when applied to lower military units, a duty assigned to an individual or unit; a task. (JP 3-0) 3. The dispatching of one or more aircraft to accomplish one particular task. (DOD Dictionary. SOURCE: JP 3-30)

nation assistance. None. (Approved for removal from the DOD Dictionary.)

neutral. In combat and combat support operations, an identity applied to a track whose characteristics, behavior, origin, or nationality indicate that it is neither supporting nor opposing friendly forces. (DOD Dictionary. SOURCE: JP 3-0)

neutrality. In international law, the attitude of impartiality during periods of war adopted by third states toward a belligerent and subsequently recognized by the belligerent, which creates rights and duties between the impartial states and the belligerent. (DOD Dictionary. SOURCE: JP 3-0)

neutralize. 1. As pertains to military operations, to render ineffective or unusable. 2. To render enemy personnel or materiel incapable of interfering with a particular operation. 3. To render safe mines, bombs, missiles, and booby traps. 4. To make harmless anything contaminated with a chemical agent. (DOD Dictionary. SOURCE: JP 3-0)

node. 1. A location in a mobility system where a movement requirement is originated, processed for onward movement, or terminated. (JP 3-17) 2. In communications and computer systems, the physical location that provides terminating, switching, and gateway access services to support information exchange. (JP 6-0) 3. An element of a system that represents a person, place, or physical thing. (DOD Dictionary. SOURCE: JP 3-0)

operation. 1. A sequence of tactical actions with a common purpose or unifying theme. (JP 1) 2. A military action or the carrying out of a strategic, operational, tactical, service, training, or administrative military mission. (DOD Dictionary. SOURCE: JP 3-0)

operational access. The ability to project military force into an operational area with sufficient freedom of action to accomplish the mission. (Approved for inclusion in the DOD Dictionary.)

operational area. An overarching term encompassing more descriptive terms (such as area of responsibility and joint operations area) for geographic areas in which military operations are conducted. Also called **OA.** (DOD Dictionary. SOURCE: JP 3-0)

operational art. The cognitive approach by commanders and staffs—supported by their skill, knowledge, experience, creativity, and judgment—to develop strategies, campaigns, and operations to organize and employ military forces by integrating ends, ways, and means. (DOD Dictionary. SOURCE: JP 3-0)

operational environment. A composite of the conditions, circumstances, and influences that affect the employment of capabilities and bear on the decisions of the commander. Also called **OE.** (DOD Dictionary. SOURCE: JP 3-0)

operational level of warfare. The level of warfare at which campaigns and major operations are planned, conducted, and sustained to achieve strategic objectives within theaters or other operational areas. (Approved for replacement of "operational level of war" and its definition in the DOD Dictionary.)

operational reach. The distance and duration across which a force can successfully employ military capabilities. (Approved for incorporation into the DOD Dictionary.)

permissive environment. Operational environment in which host country military and law enforcement agencies have control, as well as the intent and capability to assist operations that a unit intends to conduct. (Approved for incorporation into the DOD Dictionary.)

physical security. 1. That part of security concerned with physical measures designed to safeguard personnel; to prevent unauthorized access to equipment, installations,

Glossary

material, and documents; and to safeguard them against espionage, sabotage, damage, and theft. (JP 3-0) 2. In communications security, the component that results from all physical measures necessary to safeguard classified equipment, material, and documents from access thereto or observation thereof by unauthorized persons. (DOD Dictionary. SOURCE: JP 6-0)

protection. Preservation of the effectiveness and survivability of mission-related military and nonmilitary personnel, equipment, facilities, information, and infrastructure deployed or located within or outside the boundaries of a given operational area. (Approved for incorporation into the DOD Dictionary.)

protection of shipping. The use of proportionate force, when necessary for the protection of United States flag vessels and aircraft, United States citizens (whether embarked in United States or foreign vessels), and their property against unlawful violence. (Approved for incorporation into the DOD Dictionary.)

raid. An operation to temporarily seize an area in order to secure information, confuse an enemy, capture personnel or equipment, or to destroy a capability culminating with a planned withdrawal. (Approved for incorporation into the DOD Dictionary.)

risk management. The process to identify, assess, and control risks and make decisions that balance risk cost with mission benefits. Also called **RM.** (Approved for incorporation into the DOD Dictionary.)

sanction enforcement. Operations that employ coercive measures to control the movement of certain types of designated items into or out of a nation or specified area. (DOD Dictionary. SOURCE: JP 3-0)

show of force. An operation planned to demonstrate United States resolve that involves increased visibility of United States deployed forces in an attempt to defuse a specific situation that, if allowed to continue, may be detrimental to United States interests or national objectives. (Approved for incorporation into the DOD Dictionary.)

stability activities. Various military missions, tasks, and activities conducted outside the United States in coordination with other instruments of national power to maintain or reestablish a safe and secure environment, provide essential governmental services, emergency infrastructure reconstruction, and humanitarian relief. (Approved for replacement of "stability operations" and its definition in the DOD Dictionary.)

standing joint force headquarters. None. (Approved for removal from the DOD Dictionary.)

strategic level of warfare. The level of warfare at which a nation, often as a member of a group of nations, determines national or multinational (alliance or coalition) strategic security objectives and guidance, then develops and uses national resources to achieve those objectives. (Approved for the replacement of "strategic level of war" and its definition in the DOD Dictionary.)

Glossary

strategy. A prudent idea or set of ideas for employing the instruments of national power in a synchronized and integrated fashion to achieve theater, national, and/or multinational objectives. (DOD Dictionary. SOURCE: JP 3-0)

strike. An attack to damage or destroy an objective or a capability. (DOD Dictionary. SOURCE: JP 3-0)

supported commander. 1. The commander having primary responsibility for all aspects of a task assigned by the Joint Strategic Capabilities Plan or other joint planning authority. 2. In the context of joint planning, the commander who prepares operation plans or operation orders in response to requirements of the Chairman of the Joint Chiefs of Staff. 3. In the context of a support command relationship, the commander who receives assistance from another commander's force or capabilities, and who is responsible for ensuring that the supporting commander understands the assistance required. (Approved for incorporation into the DOD Dictionary.)

supporting commander. 1. A commander who provides augmentation forces or other support to a supported commander or who develops a supporting plan. 2. In the context of a support command relationship, the commander who aids, protects, complements, or sustains another commander's force, and who is responsible for providing the assistance required by the supported commander. (DOD Dictionary. SOURCE: JP 3-0)

surveillance. The systematic observation of aerospace, cyberspace, surface, or subsurface areas, places, persons, or things, by visual, aural, electronic, photographic, or other means. (Approved for incorporation into the DOD Dictionary.)

sustainment. The provision of logistics and personnel services required to maintain and prolong operations until successful mission accomplishment. (DOD Dictionary. SOURCE: JP 3-0)

system. A functionally, physically, and/or behaviorally related group of regularly interacting or interdependent elements; that group of elements forming a unified whole. (DOD Dictionary. SOURCE: JP 3-0)

tactical level of warfare. The level of warfare at which battles and engagements are planned and executed to achieve military objectives assigned to tactical units or task forces. (Approved for replacement of "tactical level of war" and its definition in the DOD Dictionary.)

targeting. The process of selecting and prioritizing targets and matching the appropriate response to them, considering operational requirements and capabilities. (DOD Dictionary. SOURCE: JP 3-0)

termination criteria. The specified standards approved by the President and/or the Secretary of Defense that must be met before a joint operation can be concluded. (DOD Dictionary. SOURCE: JP 3-0)

Glossary

terms of reference. None. (Approved for removal from the DOD Dictionary.)

theater of operations. An operational area defined by the geographic combatant commander for the conduct or support of specific military operations. Also called **TO.** (DOD Dictionary. SOURCE: JP 3-0)

theater of war. Defined by the President, Secretary of Defense, or the geographic combatant commander as the area of air, land, and water that is, or may become, directly involved in the conduct of major operations and campaigns involving combat. (DOD Dictionary. SOURCE: JP 3-0)

theater strategy. An overarching construct outlining a combatant commander's vision for integrating and synchronizing military activities and operations with the other instruments of national power in order to achieve national strategic objectives. (DOD Dictionary. SOURCE: JP 3-0)

uncertain environment. Operational environment in which host government forces, whether opposed to or receptive to operations that a unit intends to conduct, do not have totally effective control of the territory and population in the intended operational area. (DOD Dictionary. SOURCE: JP 3-0)

unity of command. The operation of all forces under a single responsible commander who has the requisite authority to direct and employ those forces in pursuit of a common purpose. (DOD Dictionary. SOURCE: JP 3-0)

weapon system. A combination of one or more weapons with all related equipment, materials, services, personnel, and means of delivery and deployment (if applicable) required for self-sufficiency. (DOD Dictionary. SOURCE: JP 3-0)

JOINT DOCTRINE PUBLICATIONS HIERARCHY

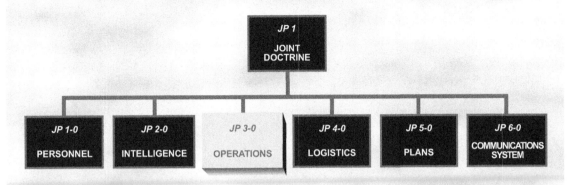

All joint publications are organized into a comprehensive hierarchy as shown in the chart above. **Joint Publication (JP) 3-0** is in the **Operations** series of joint doctrine publications. The diagram below illustrates an overview of the development process:

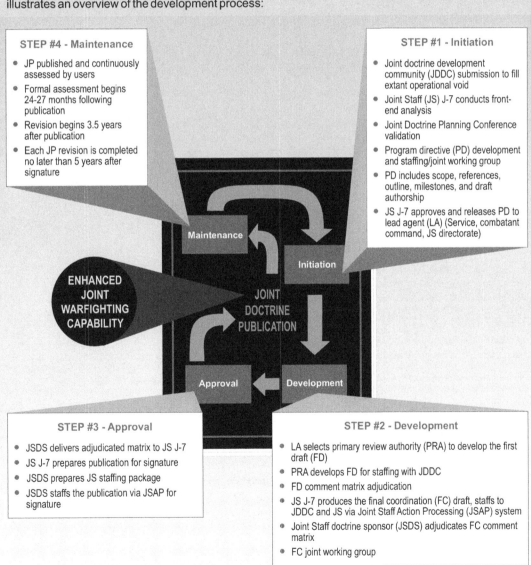

STEP #4 - Maintenance
- JP published and continuously assessed by users
- Formal assessment begins 24-27 months following publication
- Revision begins 3.5 years after publication
- Each JP revision is completed no later than 5 years after signature

STEP #1 - Initiation
- Joint doctrine development community (JDDC) submission to fill extant operational void
- Joint Staff (JS) J-7 conducts front-end analysis
- Joint Doctrine Planning Conference validation
- Program directive (PD) development and staffing/joint working group
- PD includes scope, references, outline, milestones, and draft authorship
- JS J-7 approves and releases PD to lead agent (LA) (Service, combatant command, JS directorate)

STEP #3 - Approval
- JSDS delivers adjudicated matrix to JS J-7
- JS J-7 prepares publication for signature
- JSDS prepares JS staffing package
- JSDS staffs the publication via JSAP for signature

STEP #2 - Development
- LA selects primary review authority (PRA) to develop the first draft (FD)
- PRA develops FD for staffing with JDDC
- FD comment matrix adjudication
- JS J-7 produces the final coordination (FC) draft, staffs to JDDC and JS via Joint Staff Action Processing (JSAP) system
- Joint Staff doctrine sponsor (JSDS) adjudicates FC comment matrix
- FC joint working group

Made in the USA
Las Vegas, NV
12 October 2021